Renewable Heating and Cooling

Related titles

The Biogas Handbook: Science, Production and Applications
(ISBN 978-0-85709-498-8)

Small and Micro Combined Heat and Power (CHP) Systems
(ISBN 978-1-84569-795-2)

Biomass Combustion Science, Technology and Engineering
(ISBN 978-0-85709-131-4)

Woodhead Publishing Series in Energy:
Number 89

Renewable Heating and Cooling

Technologies and Applications

Edited by

Gerhard Stryi-Hipp

ELSEVIER

AMSTERDAM • BOSTON • CAMBRIDGE • HEIDELBERG
LONDON • NEW YORK • OXFORD • PARIS • SAN DIEGO
SAN FRANCISCO • SINGAPORE • SYDNEY • TOKYO
Woodhead Publishing is an imprint of Elsevier

WP
WOODHEAD
PUBLISHING

Woodhead Publishing is an imprint of Elsevier
80 High Street, Sawston, Cambridge, CB22 3HJ, UK
225 Wyman Street, Waltham, MA 02451, USA
Langford Lane, Kidlington, OX5 1GB, UK

ISBN: 978-1-78242-213-6 (print)
ISBN: 978-1-78242-218-1 (online)

British Library Cataloguing-in-Publication Data
A catalogue record for this book is available from the British Library

Library of Congress Control Number: 2015944791

For information on all Woodhead Publishing publications visit our website at http://store.elsevier.com/

Contents

List of contributors

Ruggero Bertani Enel Green Power, Italy

Luisa F. Cabeza GREA Innovació Concurrent, Universitat de Lleida, Edifici CREA, Lleida, Spain

Roberto Fedrizzi EURAC Research, Bolzano, Italy

Stefan Hess Stellenbosch University, Stellenbosch, South Africa

Uli Jakob Dr Jakob Energy Research GmbH & Co. KG, Weinstadt, Germany; Solem Consulting, Berlin, Germany

Jan Erik Nielsen PlanEnergi, Denmark

Eduard Oró GREA Innovació Concurrent, Universitat de Lleida, Edifici CREA, Lleida, Spain

Per Alex Sørensen PlanEnergi, Denmark

Wolfram Sparber EURAC Research, Bolzano, Italy

Wolfgang Streicher Institute for Structural Engineering and Material Sciences, Unit of Energy Efficient Buildings, University of Innsbruck, Innsbruck, Austria

Gerhard Stryi-Hipp Fraunhofer Institute for Solar Energy Systems, Freiburg, Germany

Claes Tullin SP Technical Research Institute of Sweden, Borås, Sweden

Javier F. Urchueguia Technical School of Industrial Engineering, Universitat Politécnica de Valencia, Valencia, Spain

Zhiqiang Yin Tsinghua University, Beijing, China

Ruicheng Zheng Institute of Air Conditioning, China Academy of Building Research (CABR), Beijing, China

Woodhead Publishing Series in Energy

1 **Generating power at high efficiency: Combined cycle technology for sustainable energy production**
Eric Jeffs
2 **Advanced separation techniques for nuclear fuel reprocessing and radioactive waste treatment**
Edited by Kenneth L. Nash and Gregg J. Lumetta
3 **Bioalcohol production: Biochemical conversion of lignocellulosic biomass**
Edited by Keith W. Waldron
4 **Understanding and mitigating ageing in nuclear power plants: Materials and operational aspects of plant life management (PLiM)**
Edited by Philip G. Tipping
5 **Advanced power plant materials, design and technology**
Edited by Dermot Roddy
6 **Stand-alone and hybrid wind energy systems: Technology, energy storage and applications**
Edited by John K. Kaldellis
7 **Biodiesel science and technology: From soil to oil**
Jan C. J. Bart, Natale Palmeri and Stefano Cavallaro
8 **Developments and innovation in carbon dioxide (CO_2) capture and storage technology Volume 1: Carbon dioxide (CO_2) capture, transport and industrial applications**
Edited by M. Mercedes Maroto-Valer
9 **Geological repository systems for safe disposal of spent nuclear fuels and radioactive waste**
Edited by Joonhong Ahn and Michael J. Apted
10 **Wind energy systems: Optimising design and construction for safe and reliable operation**
Edited by John D. Sørensen and Jens N. Sørensen
11 **Solid oxide fuel cell technology: Principles, performance and operations**
Kevin Huang and John Bannister Goodenough
12 **Handbook of advanced radioactive waste conditioning technologies**
Edited by Michael I. Ojovan
13 **Membranes for clean and renewable power applications**
Edited by Annarosa Gugliuzza and Angelo Basile
14 **Materials for energy efficiency and thermal comfort in buildings**
Edited by Matthew R. Hall
15 **Handbook of biofuels production: Processes and technologies**
Edited by Rafael Luque, Juan Campelo and James Clark
16 **Developments and innovation in carbon dioxide (CO_2) capture and storage technology Volume 2: Carbon dioxide (CO_2) storage and utilisation**
Edited by M. Mercedes Maroto-Valer

17 **Oxy-fuel combustion for power generation and carbon dioxide (CO_2) capture**
Edited by Ligang Zheng
18 **Small and micro combined heat and power (CHP) systems: Advanced design, performance, materials and applications**
Edited by Robert Beith
19 **Advances in clean hydrocarbon fuel processing: Science and technology**
Edited by M. Rashid Khan
20 **Modern gas turbine systems: High efficiency, low emission, fuel flexible power generation**
Edited by Peter Jansohn
21 **Concentrating solar power technology: Principles, developments and applications**
Edited by Keith Lovegrove and Wes Stein
22 **Nuclear corrosion science and engineering**
Edited by Damien Féron
23 **Power plant life management and performance improvement**
Edited by John E. Oakey
24 **Electrical drives for direct drive renewable energy systems**
Edited by Markus Mueller and Henk Polinder
25 **Advanced membrane science and technology for sustainable energy and environmental applications**
Edited by Angelo Basile and Suzana Pereira Nunes
26 **Irradiation embrittlement of reactor pressure vessels (RPVs) in nuclear power plants**
Edited by Naoki Soneda
27 **High temperature superconductors (HTS) for energy applications**
Edited by Ziad Melhem
28 **Infrastructure and methodologies for the justification of nuclear power programmes**
Edited by Agustín Alonso
29 **Waste to energy conversion technology**
Edited by Naomi B. Klinghoffer and Marco J. Castaldi
30 **Polymer electrolyte membrane and direct methanol fuel cell technology Volume 1: Fundamentals and performance of low temperature fuel cells**
Edited by Christoph Hartnig and Christina Roth
31 **Polymer electrolyte membrane and direct methanol fuel cell technology Volume 2: *In situ* characterization techniques for low temperature fuel cells**
Edited by Christoph Hartnig and Christina Roth
32 **Combined cycle systems for near-zero emission power generation**
Edited by Ashok D. Rao
33 **Modern earth buildings: Materials, engineering, construction and applications**
Edited by Matthew R. Hall, Rick Lindsay and Meror Krayenhoff
34 **Metropolitan sustainability: Understanding and improving the urban environment**
Edited by Frank Zeman
35 **Functional materials for sustainable energy applications**
Edited by John A. Kilner, Stephen J. Skinner, Stuart J. C. Irvine and Peter P. Edwards
36 **Nuclear decommissioning: Planning, execution and international experience**
Edited by Michele Laraia
37 **Nuclear fuel cycle science and engineering**
Edited by Ian Crossland
38 **Electricity transmission, distribution and storage systems**
Edited by Ziad Melhem

39 **Advances in biodiesel production: Processes and technologies**
Edited by Rafael Luque and Juan A. Melero
40 **Biomass combustion science, technology and engineering**
Edited by Lasse Rosendahl
41 **Ultra-supercritical coal power plants: Materials, technologies and optimisation**
Edited by Dongke Zhang
42 **Radionuclide behaviour in the natural environment: Science, implications and lessons for the nuclear industry**
Edited by Christophe Poinssot and Horst Geckeis
43 **Calcium and chemical looping technology for power generation and carbon dioxide (CO_2) capture: Solid oxygen- and CO_2-carriers**
Paul Fennell and E. J. Anthony
44 **Materials' ageing and degradation in light water reactors: Mechanisms, and management**
Edited by K. L. Murty
45 **Structural alloys for power plants: Operational challenges and high-temperature materials**
Edited by Amir Shirzadi and Susan Jackson
46 **Biolubricants: Science and technology**
Jan C. J. Bart, Emanuele Gucciardi and Stefano Cavallaro
47 **Advances in wind turbine blade design and materials**
Edited by Povl Brøndsted and Rogier P. L. Nijssen
48 **Radioactive waste management and contaminated site clean-up: Processes, technologies and international experience**
Edited by William E. Lee, Michael I. Ojovan, Carol M. Jantzen
49 **Probabilistic safety assessment for optimum nuclear power plant life management (PLiM): Theory and application of reliability analysis methods for major power plant components**
Gennadij V. Arkadov, Alexander F. Getman and Andrei N. Rodionov
50 **The coal handbook: Towards cleaner production Volume 1: Coal production**
Edited by Dave Osborne
51 **The coal handbook: Towards cleaner production Volume 2: Coal utilisation**
Edited by Dave Osborne
52 **The biogas handbook: Science, production and applications**
Edited by Arthur Wellinger, Jerry Murphy and David Baxter
53 **Advances in biorefineries: Biomass and waste supply chain exploitation**
Edited by Keith Waldron
54 **Geological storage of carbon dioxide (CO_2): Geoscience, technologies, environmental aspects and legal frameworks**
Edited by Jon Gluyas and Simon Mathias
55 **Handbook of membrane reactors Volume 1: Fundamental materials science, design and optimisation**
Edited by Angelo Basile
56 **Handbook of membrane reactors Volume 2: Reactor types and industrial applications**
Edited by Angelo Basile
57 **Alternative fuels and advanced vehicle technologies for improved environmental performance: Towards zero carbon transportation**
Edited by Richard Folkson

58 **Handbook of microalgal bioprocess engineering**
Christopher Lan and Bei Wang
59 **Fluidized bed technologies for near-zero emission combustion and gasification**
Edited by Fabrizio Scala
60 **Managing nuclear projects: A comprehensive management resource**
Edited by Jas Devgun
61 **Handbook of Process Integration (PI): Minimisation of energy and water use, waste and emissions**
Edited by Jiří J. Klemeš
62 **Coal power plant materials and life assessment**
Edited by Ahmed Shibli
63 **Advances in hydrogen production, storage and distribution**
Edited by Ahmed Basile and Adolfo Iulianelli
64 **Handbook of small modular nuclear reactors**
Edited by Mario D. Carelli and Dan T. Ingersoll
65 **Superconductors in the power grid: Materials and applications**
Edited by Christopher Rey
66 **Advances in thermal energy storage systems: Methods and applications**
Edited by Luisa F. Cabeza
67 **Advances in batteries for medium and large-scale energy storage**
Edited by Chris Menictas, Maria Skyllas-Kazacos and Tuti Mariana Lim
68 **Palladium membrane technology for hydrogen production, carbon capture and other applications**
Edited by Aggelos Doukelis, Kyriakos Panopoulos, Antonios Koumanakos and Emmanouil Kakaras
69 **Gasification for synthetic fuel production: Fundamentals, processes and applications**
Edited by Rafael Luque and James G. Speight
70 **Renewable heating and cooling: Technologies and applications**
Edited by Gerhard Stryi-Hipp
71 **Environmental remediation and restoration of contaminated nuclear and NORM sites**
Edited by Leo van Velzen
72 **Eco-friendly innovation in electricity networks**
Edited by Jean-Luc Bessede
73 **The 2011 Fukushima nuclear power plant accident: How and why it happened**
Yotaro Hatamura, Seiji Abe, Masao Fuchigami and Naoto Kasahara. Translated by Kenji Iino
74 **Lignocellulose biorefinery engineering: Principles and applications**
Hongzhang Chen
75 **Advances in membrane technologies for water treatment: Materials, processes and applications**
Edited by Angelo Basile, Alfredo Cassano and Navin Rastogi
76 **Membrane reactors for energy applications and basic chemical production**
Edited by Angelo Basile, Luisa Di Paola, Faisal Hai and Vincenzo Piemonte
77 **Pervaporation, vapour permeation and membrane distillation: Principles and applications**
Edited by Angelo Basile, Alberto Figoli and Mohamed Khayet

78 **Safe and secure transport and storage of radioactive materials**
Edited by Ken Sorenson
79 **Reprocessing and recycling of spent nuclear fuel**
Edited by Robin Taylor
80 **Advances in battery technologies for electric vehicles**
Edited by Bruno Scrosati, Jürgen Garche and Werner Tillmetz
81 **Rechargeable lithium batteries: From fundamentals to applications**
Edited by Alejandro A. Franco
82 **Calcium and chemical looping technology for power generation and carbon dioxide (CO_2) capture**
Edited by Paul Fennell and Ben Anthony
83 **Compendium of Hydrogen Energy Volume 1: Hydrogen Production and Purificiation**
Edited by Velu Subramani, Angelo Basile and T. Nejat Veziroglu
84 **Compendium of Hydrogen Energy Volume 2: Hydrogen Storage, Transmission, Transportation and Infrastructure**
Edited by Ram Gupta, Angelo Basile and T. Nejat Veziroglu
85 **Compendium of Hydrogen Energy Volume 3: Hydrogen Energy Conversion**
Edited by Frano Barbir, Angelo Basile and T. Nejat Veziroglu
86 **Compendium of Hydrogen Energy Volume 4: Hydrogen Use, Safety and the Hydrogen Economy**
Edited by Michael Ball, Angelo Basile and T. Nejat Veziroglu
87 **Advanced district heating and cooling (DHC) systems**
Edited by Robin Wiltshire
88 **Microbial Electrochemical and Fuel Cells: Fundamentals and Applications**
Edited by Keith Scott and Eileen Hao Yu
89 **Renewable Heating and Cooling: Technologies and Applications**
Edited by Gerhard Stryi-Hipp
90 **Small Modular Reactors: Nuclear Power Fad or Future?**
Edited by Daniel T. Ingersoll
91 **Fuel Flexible Energy Generation: Solid, Liquid and Gaseous Fuels**
Edited by John Oakey

Introduction to renewable heating and cooling

1

Gerhard Stryi-Hipp
Fraunhofer Institute for Solar Energy Systems, Freiburg, Germany

Our present energy system, constructed during the past century, will be radically transformed in the twenty-first century according the Group of Seven (G7) industrial nations, which ended their summit in June 2015 by talking of the "decarbonization of the global economy over the course of this century." They promised to cut greenhouse-gas emissions by the "upper end" of a range between 40 and 70% of 2010 levels by 2050 (ECO, 2015).

Usually politicians mean only electricity when they talk about "energy," besides the fact that the production of heat accounts for around one-third of global energy-related carbon dioxide (CO_2) emissions and half of the final energy demand (IEA, 2014). One reason for that is the traditional centralized electricity system in which most of the generators and consumers are connected to the electrical grid, at least in industrialized countries. The collectively used electrical grid and the need for a secure electrical supply force the governments to regulate and control the electrical system. Another reason is that building up the electrical system by electrifying street lighting, factories, households, rural areas, and railways was controlled from the beginning in the 1880s on by the individual states (NEA, 2015).

In contrast, heating is traditionally decentralized because humans learned to use, preserve, and make fire more than 30,000 years ago. Heat for cooking, domestic hot water, and space heating, as well as for industrial processes is traditionally produced individually in buildings or close to the places of use. An important difference to electricity is that there is the possibility, but not the need, of a centralized heat supply by using a heating grid. Centralized heat supply is well known and was used for the first time by the Romans 3000 years ago. The first district heating system to supply a city was already built in 1332 in the French village of Chaudes-Aigues, where 82 °C thermal water was delivered to buildings by wooden pipes (Geo, 2015). However, only a small share of heat demand is supplied by district heating systems today.

In recent years, the awareness of governments and institutions of the importance of the heating sector is rising, due to strong fossil fuel price increases during the past decade, the growing fossil fuel import dependency of most countries, and the growing need of decarbonization, which is impossible without a strong contribution of the heating sector. In 2011, final energy heat consumption accounted for 171.5 EJ globally, which corresponds to 50.6% of the total final energy use of 339 EJ, whereas 65.1 EJ (19.2%) was used as electricity and 102.4 EJ (30.2%) in the transport sector (Figure 1.1). Global energy use for heat of 129 EJ (75.2%) is met with fossil fuels. Forty percent of

Renewable Heating and Cooling. http://dx.doi.org/10.1016/B978-1-78242-213-6.00001-1

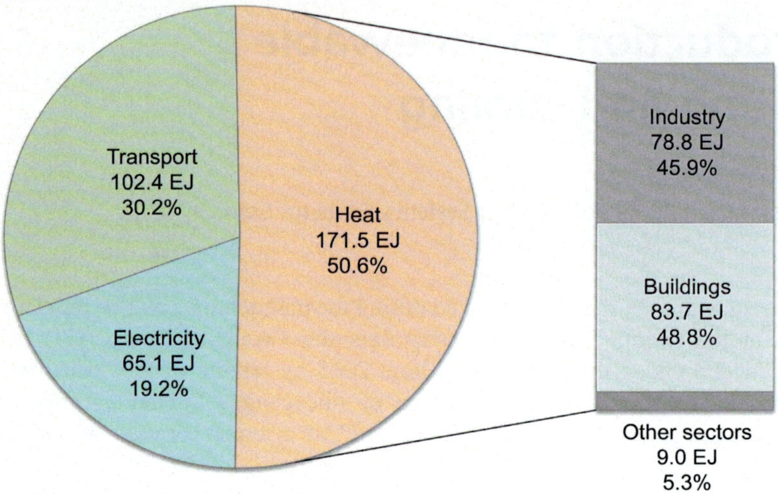

Figure 1.1 Final energy demand globally, distribution of heat use by sectors.
Source: IEA (2014).

the primary energy supply of natural gas, as well as 20% each of oil and coal are being used for heat production, with important impacts on energy security (IEA, 2014).

Renewable energy accounts for 43% (36 EJ) of total energy use for heat in buildings. However, most of this comes from the traditional use of biomass for cooking and space heating in developing and emerging economies. Such fuel use is usually unsustainable and is a cause of deforestation and health problems linked to indoor smoke pollution, among other disadvantages. Only 4 EJ are currently produced by more sustainable renewable energy technologies. This so-called modern bioenergy makes the largest contribution (3 EJ), whereas the use of solar thermal (0.7 EJ) and geothermal energy (0.3 EJ) for heat in buildings is small by comparison. In the industry sector, which consumes 78.8 EJ for heating, renewable energy use for heat accounts for 10% of the total, of which 99% is bioenergy based. Because buildings use 83.7 EJ, industry 78.8 EJ, and other sectors 9.0 EJ for heat, only 7% (12 EJ) of the final energy consumption for heat is provided by sustainable renewable energy sources, of which 92% is bioenergy based (IEA, 2014).

Today's use of sustainable renewable energy for heating and cooling is very low in comparison to its high potential. According to the European Technology Platform for Renewable Heating and Cooling, renewable energy sources could meet the overall heating and cooling demand in Europe already by 2040 (RHC, 2013). The strengths of renewable energy sources for heating are sustainability and local availability. They are available in each country in an individual country mix and can be harvested close to the point of heat demand. In addition, heating technologies differ a lot from region to region and are adapted to local conditions. Unfortunately, the local diversity of sources and of technologies used hamper the deployment of renewable heating and cooling (RHC) technologies, because it is impossible to build up RHC mass markets with standardized solutions to bring down costs and allow a fast diffusion of the technology. This is also the reason why simple policy instruments are not sufficient to stimulate market deployment of the manifold RHC technologies.

There are a large number of reasons why the deployment of RHC technologies is much lower than the market success of renewable electrical technologies. Because most heating systems are not connected to a heating network, thermal storage tanks are needed to compensate the mismatch between heat generation and consumption, especially if the RHC source is not storable as it is, for example, for the case in solar thermal systems for space heating at higher latitudes. In contradiction, each kWh of generated electricity by renewable energy can be fed into the grid and be directly used. This is at least the case in the beginning of the deployment phase with a low share of renewable electricity generation. RHC systems usually need storage capacities, which lead to additional costs in comparison with electrical generation systems.

There is a high diversity of heating and cooling technologies due to the high diversity of heat demand types according size and temperature. Heat is used for cooking, domestic hot water heating, space heating, and industrial processes in a broad temperature range from 30 °C up to above 1000 °C. The demand differs a lot between countries and regions depending on local climates. Space heating is needed only in cold and moderate climates, and cooling is obligatory in hot climates if people can afford it. But heat demand is not only affected by climatic conditions but also by the typical thermal comfort based on local traditions and cultural distinctions. It should be noted in addition that the level of heat demand is the result of the expected thermal comfort and the efficiency standard of a building, but the primary energy demand is additionally determined by the efficiency of the heating technology used.

Heating and cooling technologies depend also on the size of the buildings, the number of people, or the size of the industry supplied with heat and cold. Heating boilers for single-family homes and big combined heat and power plants for district heating systems, as well as split systems for air conditioning and big absorption chillers differ greatly. In addition, the type of distribution system influences the heat and cold generation technology. Single-room heating with stoves placed in the rooms to be heated is mainly used in small buildings. Water-based central heating systems for apartments, small and large apartment buildings, or building complexes require central boilers with adapted capacities. In air heating systems specific air heaters or heat exchangers are used. District heating systems can be supplied by large combined heat and power plants in combination with peak load boilers. And for all applications different types of solar thermal, biomass, geothermal, and heat pump systems are available as well.

RHC technologies must not only be adapted to the demand side and its diversity of heat distribution technologies, but also to the resource side, because solar thermal systems depend on the intensity and variability of solar radiation and geothermal systems on the local conditions of the ground or the water. Biomass benefits from its flexibility because it can be transported and stored; however, costs and CO_2 emissions are increasing with the distance between the place of growth and the place of use of biomass, and therefore the availability of biomass in the region around the user is a limiting factor for the use of biomass.

The heating and cooling sector shows also a higher diversity on investors and type of investments in comparison with the electricity sector. Building owners and companies investing in stoves, boilers, or electrical heaters and buying primary energy sources to generate heat. Only a minor share is buying heat from district heating systems. The investment motivation and therefore investment decisions differ a lot with the types of building owners, housing companies, and investors. Private residential home owners, private landlords, cooperative and commercial housing companies, as well as enterprises and district heating companies are characterized by different investment needs and investment criteria. Therefore, several separated heating and cooling technology markets result and must be addressed individually by the manufacturers and distributors.

The specific conditions of the heating and cooling markets described are significant challenges for RHC technologies to enter these markets. However, the main barrier is the price of fossil fuels and the often missing competitiveness. Usually RHC technologies have higher investment costs, no or low fuel costs, and lower maintenance costs than traditional heating technologies. To compare the costs, the average costs per kWh over the lifetime of the system are calculated based on assumptions of fuel cost development. Even if the renewable heating costs over the lifetime are lower than from fossil fuels based on the assumption of a fossil fuel price increase rate, the higher investment costs and the high uncertainty of the future fuel price development are a significant barrier for a lot of investors. This barrier is further increased by the huge amount of subsidies provided for fossil fuels globally.

However, there are strong drivers for the transformation from fossil fuels to RHC technologies: the need of decarbonization to limit the temperature increase caused by climate change and the growing fossil fuel import dependency in most of the countries globally. Today many countries spend a significant share of their state budget for fuel imports. RHC technologies offer the opportunity to replace them and create local added value and jobs. Because RHC technologies have a huge potential for technical improvements, it can be expected that the competitiveness of RHC technologies will continuously grow in the future by lower costs and increased attractiveness.

Against this background it is necessary to intensify research and development as well as the political support for global market deployment of RHC technologies. Based on a deeper understanding of the specific technical and nontechnical challenges of the different RHC technologies, fields of innovation can be identified. This book aims to support this process by providing a sound overview on the status and the potential of biomass, solar thermal, geothermal, and heat pump technologies, as well as on hybrid systems. I wish you an interesting reading and many new insights in the exciting RHC technologies and hope that this book will contribute to unlock the potential of the RHC technologies.

References

ECO, 2015 The Economist, Sort of, Why G7 is talking about carbonization, June 10, 2015.

RHC, 2013 RHC-platform, Strategic Research and Innovation Agenda for Renewable Heating and Cooling, Brussels, 2013.

GEO, 2015 www.geothermie.ch/index.php?p=keyfacts_thermalWater.
NEA, 2015 National Academy of Engineering, www.greatachievements.org/?id=2988.
IEA, 2014 IEA, Heating without global warming, Market developments and Policy Considerations for Renewable Heat, Paris, 2014.

Part One

Key technologies for renewable heating and cooling and their applications

Solar thermal technologies for domestic hot water preparation and space heating

2

Wolfgang Streicher
Institute for Structural Engineering and Material Sciences, Unit of Energy Efficient Buildings, University of Innsbruck, Innsbruck, Austria

2.1 Introduction

Solar thermal energy has been used for thousands of years in openings and the windows of buildings as so-called passive solar energy (see Figure 2.1). Different solar incident angles in summer and winter are used in the design and orientation of buildings to get maximum solar gain in winter and minimum solar gain in summer. This is achieved by a southern orientation, an overhang that gives shading in summer but lets in the sun in winter, and the thermal mass of a building that dampens the internal day/night temperature variation.

The application of this concept can also save heating and cooling energy in today's buildings, but this knowledge seems to have been forgotten by many architects.

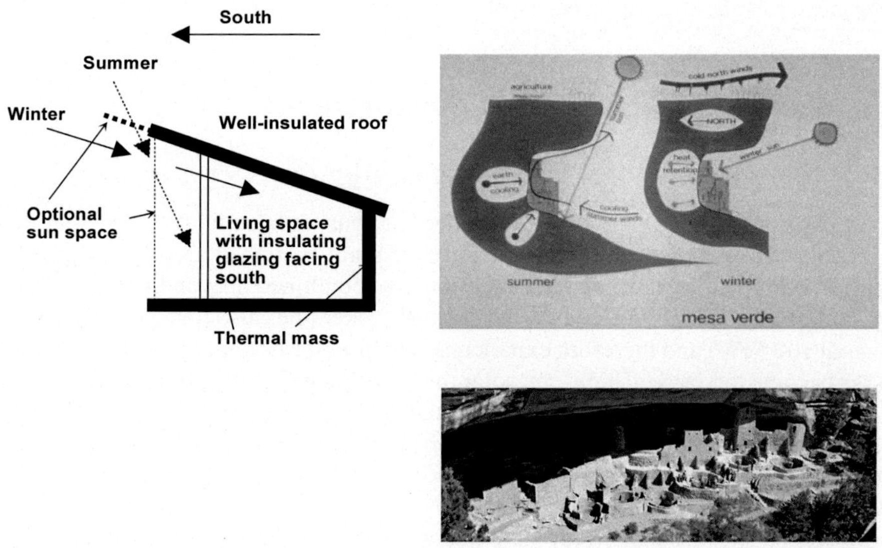

Figure 2.1 Left side: concept of passive solar energy use of the so-called SOCRATES house (around 500 B.C.) (Streicher, 2015). Right side: passive solar energy use in the pueblo of Mesa Verde, Colorado, USA, 700 A.D.

Renewable Heating and Cooling. http://dx.doi.org/10.1016/B978-1-78242-213-6.00002-3

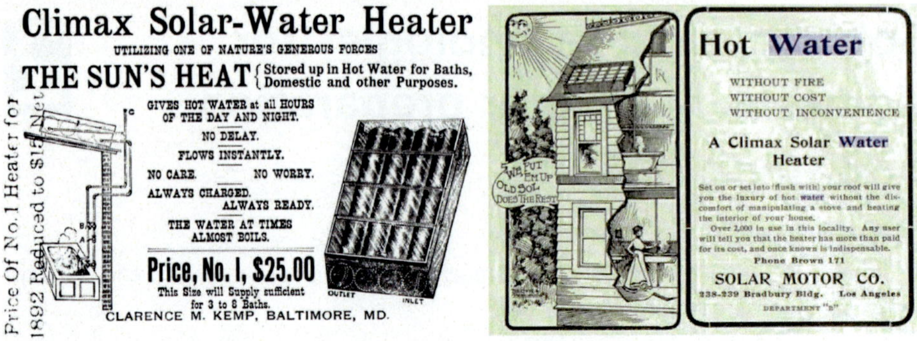

Figure 2.2 An advertisement for a solar water heater dated 1896 and 1902.
Source: (left) http://energyblog.nationalgeographic.com/2013/09/23/seven-of-the-greatest-solar-stories-over-the-millennia/; (right) http://en.wikipedia.org/wiki/Solar_water_heating#History, accessed 01.03.15.

Active solar energy use came into operation around the year 1900, when solar water roof heaters were marketed (see Figure 2.2). A real market became established after the first oil crisis at the end of the 1970s. Today, different applications ranging from the heating of swimming pools, domestic hot water (DHW), space heating (SH), so-called combisystems that combine DHW and SH, district heat, process heat, solar thermal cooling, and electricity production by solar thermal power plants have worldwide markets. Since 2005 photovoltaic (PV) has become so inexpensive that it is now a competitor to solar thermal in many of these applications.

2.2 Potentials and market development

2.2.1 Potentials for solar thermal energy use

Apart from the different aforementioned applications, the potential use of solar thermal energy is restricted by the fact that heat cannot be transported long distances. District heating networks may have heat sources (combined heat and power stations) 30–50 km away from the network, but such hot water lines also transport high power (several 100 MW) and therefore experience low losses. For lower power, the specific losses become too large. Therefore, solar thermal heat has to be produced close to the user (e.g., in the same building). The potential areas are roofs and façades with southeast to southwest orientation (in the Northern Hemisphere), or flat roofs of buildings. For district heating systems open land close to the district heating network may be used.

Moreover, no seasonal storage for hot water is available on the market. This is partly due to the fact that seasonal storage only "earns" money once a year, whereas daily storage "earns" money 365 times a year. Additionally, even very large well-insulated storages of 10,000 m^3 or more lose too much temperature over the storing months for direct use. Most existing test stores have been equipped with a heat pump for heat extraction. Seasonal storage would be needed because the solar radiation in Middle Europe on a horizontal surface is about five times higher in summer than in winter.

These two effects reduce the potential of solar thermal systems. If a solar thermal plant is designed for high yearly fractional saving (this is the ratio of saved conventional energy by the solar plant compared to a conventional plant), in summer there is always an unusable surplus of energy that leads to higher cost per kWh of the plant. If the plant is designed to cover only the summer load, the fractional solar saving in winter is very small although the specific cost per kWh is reduced.

Economically possible solar potentials range from 6% to over 50% of the heat demand of buildings and industry depending on the boundary conditions (Kaltschmitt, Streicher, & Wiese, 2013).

2.2.2 International market development

International market data are collected annually by Mauthner and Weiss (2014) for the Solar Heating and Cooling program of the International Energy Agency. These data are used in this subsection.

The market of solar thermal energy is dominated by China, where about 74% of all flat-plate and evacuated-tube collectors are in operation, followed by Europe with 16% (see Figure 2.3). In total, 250 megawatt thermal (MWth) heating power has been installed worldwide in 2012. The total number of jobs in the field of production, installation, and maintenance is estimated at 460,000 worldwide.

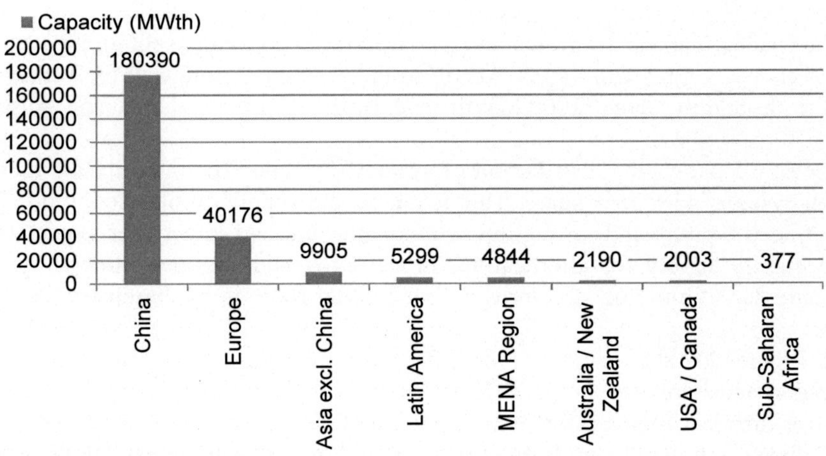

Figure 2.3 Total capacity of glazed flat-plate and evacuated-tube collectors in operation by economic region through 2012.
Values from Mauthner and Weiss (2014).

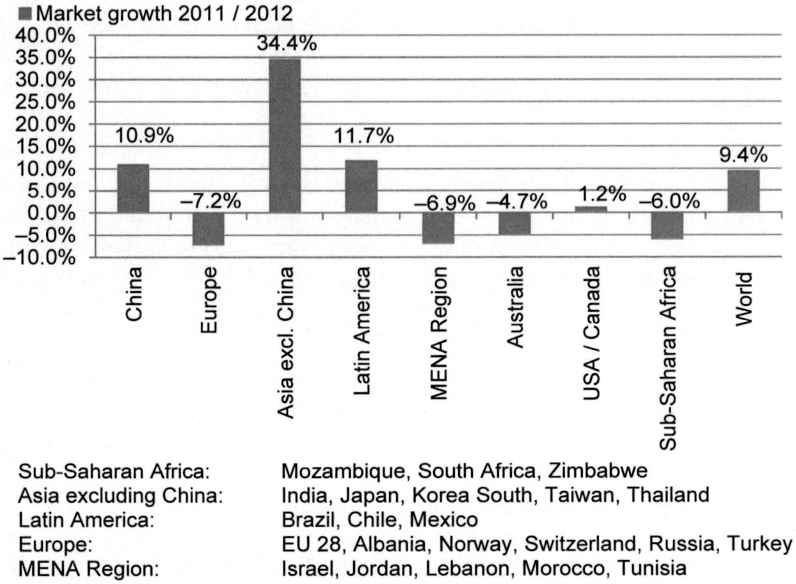

Figure 2.4 Market growth of newly installed capacity between 2011 and 2012 by economic region market development.
Values from Mauthner and Weiss (2014).

The highest growth rate in 2012 was achieved in Asia, excluding China, with 34.4% followed by Latin America. The growth rate of China still was 10.9%. In Europe, the Middle East–North Africa (MENA) region, Australia, and Sub-Saharan Africa the market was declining between 5 and 7% (see Figure 2.4). This clearly points to future markets apart from China and the current difficult situation of solar thermal use in Europe. Possible explanations for this are given in the next subsection.

The leading markets for glazed and unglazed water collectors split up by countries are, of course, China with 44,000 MWth installed in 2012, followed by Turkey and India with slightly above 1000 MWth, and Brazil, Germany, USA, and Australia with 650–800 MWth.

Figure 2.5 shows the distribution of systems by type. The majority are thermosiphon systems with 75% share. This is due to the simplicity of installation. They do not need a pump and, in regions without freezing and lower user expectations, no electricity supply for afterheating. In the USA and Europe, pumped systems dominate due to building structures with the boiler room in the basement and tilted roofs.

In Figure 2.6 the market is described by the applications for which solar thermal systems are used in the different regions. It shows completely different pictures for different regions. Swimming-pool applications dominate in USA/Canada, Australia/New Zealand, and Sub-Saharan Africa. For all other areas DHW production is the main application. DHW for single-family houses has the highest share

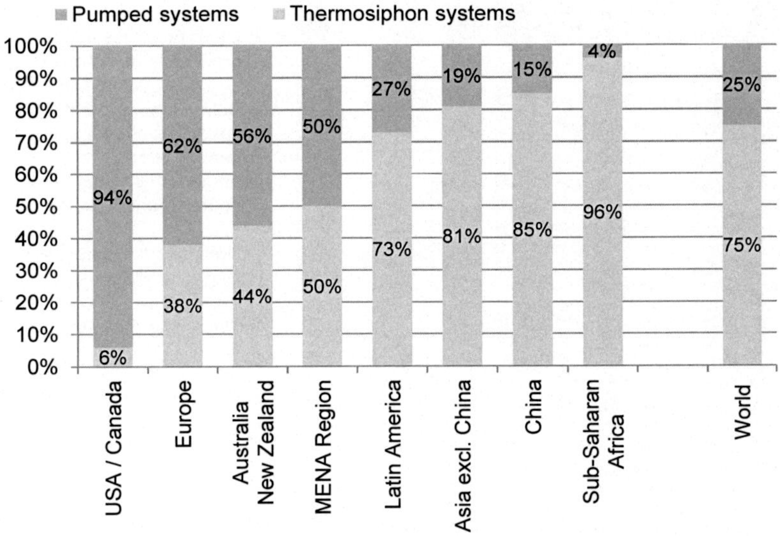

Sub-Saharan Africa: Namibia, South Africa
Asia excluding China: India, Japan, Korea South, Taiwan, Thailand
Latin America: Brazil, Chile
Europe: EU 27, Albania, Norway, Switzerland, Turkey
MENA Region: Israel, Jordan, Lebanon, Tunisia

Figure 2.5 Distribution by type of system for the total installed glazed water collector capacity in operation through 2012.
Values from Mauthner and Weiss (2014).

in China with 90% and Asia with 82% followed by Europe with 62% and Latin America with 54%. In the MENA region only, DHW systems are built whereby large-scale DHW installations have an even higher share compared to DHW systems for single-family houses. Solar combisystems are systems combining SH and DHW production, which are a European specialty that is very slowly gaining market share in other regions.

Other new applications, such as the production of low-temperature industrial process heat or solar thermal cooling systems, are currently in the research and pilot plant stage. Several industrial plants have been built in the past years for, e.g., breweries, washing purposes, or food drying. About 1050 solar cooling plants have been built by 2012. Nevertheless, the current market is very small. Industrial processes may gain higher interest for solar thermal applications in the future.

2.2.3 Solar thermal versus PV applications

One of the big challenges of solar thermal applications is the massive price reduction of PV plants since 2005. Although the price for solar thermal systems did not significantly drop between 2001 and 2013, the price for PV has dropped by 70% between 2009 and 2014 (see Figures 2.7 and 2.8). Current installations of large

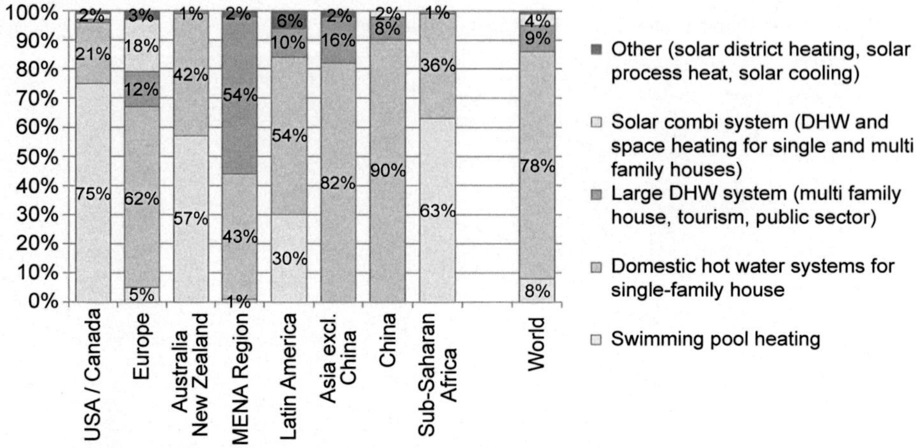

DHW domestic hot water
Sub-Saharan Africa: Namibia, South Africa
Asia excluding China: India, Japan, Korea South, Taiwan
Latin America: Brazil, Chile, Mexico
Europe: EU 18, Albania, Norway, Switzerland, Russia, Turkey
MENA Region: Israel, Jordan, Lebanon, Tunisia

Figure 2.6 Distribution of solar thermal systems by application for the total installed water collector capacity by economic region in operation through 2012.
Values from Mauthner and Weiss (2014).

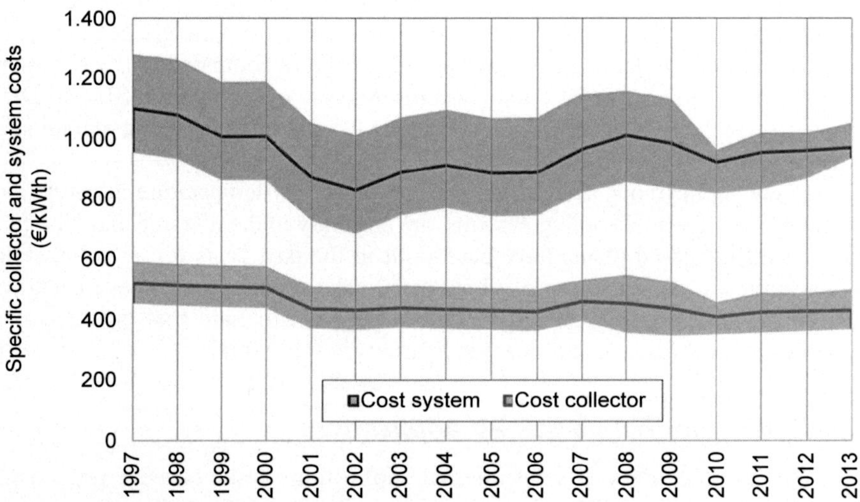

Figure 2.7 Price development of solar thermal plants for hot water production in Austria (inflation adjusted based on 2013).
Source: AEE-INTEC in Biermayr et al. (2014).

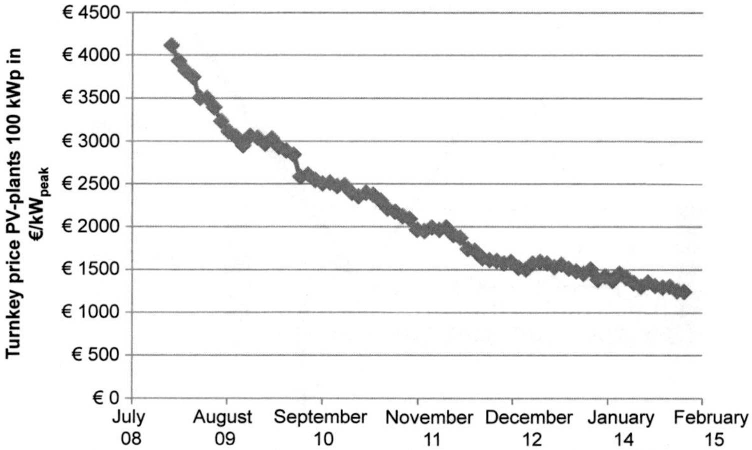

Figure 2.8 End-user price development of 100 kW_{peak} turnkey photovoltaic plants in Germany in €/W_{peak}.
Values from http://www.photovoltaik-guide.de/pv-preisindex, downloaded 02/2015.

PV plants and small solar thermal plants for DHW production have nearly the same price per kW installed. PV is able to fulfill the same applications as solar thermal, but excess solar energy can be fed as electricity into the grid; so excess solar energy is not lost as in solar thermal systems. Of course, PV has about 10% lower seasonal efficiency than solar thermal systems with about 30% (depending strongly on the fractional solar saving of the installation, see Section 4.3). On the other hand, if the electricity from PV is used in heat pumps with a seasonal performance factor of 3, the efficiency from sun to heat is the same for both technologies, and equal areas are needed for the same applications. With today's prices even direct resistance heating without a heat pump seems to be economic compared to solar thermal applications under specific boundary conditions. Additionally, installation of PV systems is often far simpler than the hydraulics of solar thermal systems (ref. to Sections 3.2−3.6).

The great challenge for solar thermal systems is, therefore, on the one hand to reduce the price of the systems and on the other to make systems simpler for installation and maintenance.

2.3 Components of solar thermal collector systems

2.3.1 Collectors

The physical principle of a solar thermal collector is a black absorber plate that is heated up by absorbed solar radiation. Figure 2.9 shows this principle for a flat plate collector. The heat has to be driven by heat conduction to the heat carrier that, most

η_0 (−)	Conversion factor or optical losses; efficiency when $\overline{T}_{abs} = T_{amb}$ ($\eta_0 \approx \tau_{cov} \cdot \alpha_{abs}$)
a_1 (W/m^2 K)	Overall heat transfer coefficient related to $\overline{T}_{abs} = T_{amb}$. This factor is normally given at 0 and sometimes at 4 m/s wind speed
a_2 (W/m^2 K^2)	Quadratic term; this is a black box approach for the nonlinear radiation losses and the temperature dependency of the heat transfer coefficient

The European standard EN 12975-2 (2006) describes how these curves are derived by measurements. It has to be reported whether the collector efficiency values a_1 and a_2 may refer to either the gross area, the aperture area, or the absorber area of the collector (Figure 2.10).

Figure 2.11 shows the efficiency curves for different typical collector types related to collector gross area and the range of specific temperature differences $(\overline{T}_{abs} - T_{amb})/G_g$ for several applications for Middle European climate. Clearly the dependency of the heat losses on the specific temperature difference can be seen. Additionally, at zero temperature difference, the optical losses η_0 due to reflection and absorption of the glass cover and the reflection of the absorber can be seen. Swimming-pool collectors have the highest efficiency of all collector types at zero temperature difference due to the missing glass cover. Vacuum-tube collectors have the lowest η_0 because the ratio of absorber to gross area is quite small (Figure 2.10).

As swimming-pool collectors have no glass cover they have high convective losses, and the efficiency curve drops fast with increasing temperature difference. This is an ideal curve for swimming pools, because the swimming-pool temperature, and therefore the absorber temperature, is often near or below the ambient temperature.

Flat-plate collectors and, even more, flat-plate collectors with a selective surface on the absorber have slightly lower efficiencies compared to swimming-pool collectors at very small temperature differences but far higher efficiencies at higher temperature differences. They are best suited for DHW and solar combisystems. Vacuum tube collectors are best suited for high-temperature differences

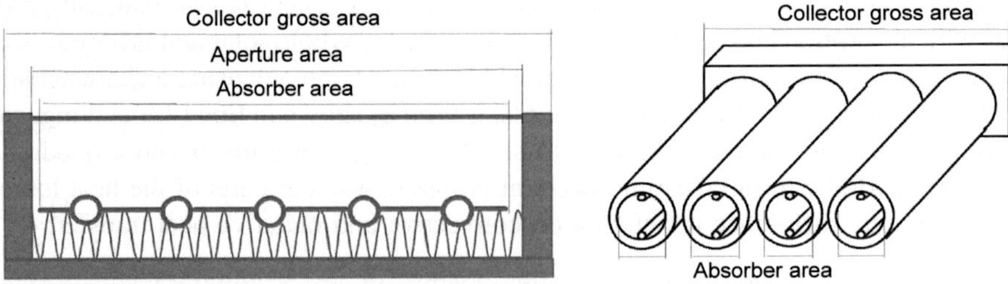

Figure 2.10 Definition of the different collector areas for a flat-plate and a vacuum-tube collector (Streicher, 2015).

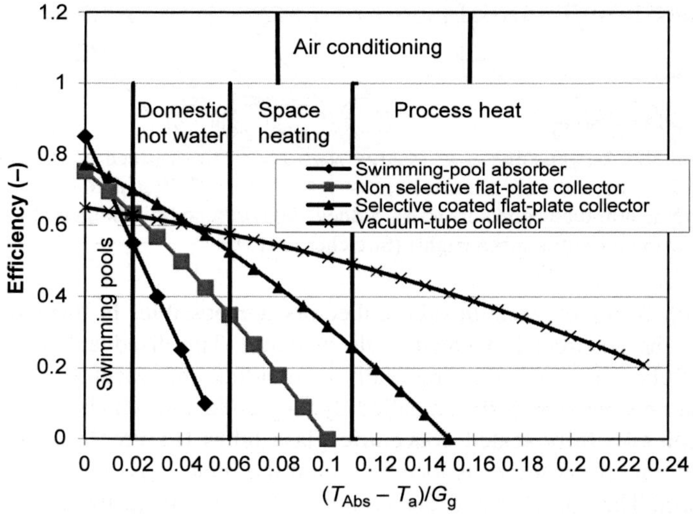

Figure 2.11 Collector characteristic efficiency curves for different typical collector types related to collector gross area and temperature ranges of typical applications (Streicher, 2015).

between absorber and ambient temperature, e.g., for solar air conditioning or solar process heat.

In addition, the maximum temperature at stagnation can be calculated from the curves in Figure 2.11. Stagnation means that no useful heat is taken out and the efficiency is therefore zero. With the x-value at the crossing of the curve with the x-axis the maximum absorber, temperature can be calculated by setting the ambient temperature and the irradiance to worst cases, e.g., $T_a = 30\ °C$ and $G_g = 1000\ W/m^2$. For the selectively coated flat-plate collector, the crossing of the x-axis is at $0.15 = (\overline{T}_{abs} - T_{amb})/G_g$. With the above assumptions $\overline{T}_{abs}$ becomes 180 °C, which is reasonable for good flat-plate collectors. Of course, all components of the collector have to withstand such high temperatures, and care has to be taken to deal with the evaporation of the collector fluid (see Section 3.4).

Figure 2.12 shows examples for swimming-pool and flat-plate absorbers. Swimming-pool collectors operate at temperatures between 20 and 30 °C, and the stagnation temperature is not above 80 °C Therefore, these absorbers are often made from polymers. As polymers have a low thermal conductivity these absorbers have very short or even no fins but a high number of parallel tubes. Flat-plate collectors have been described earlier (see Figure 2.9).

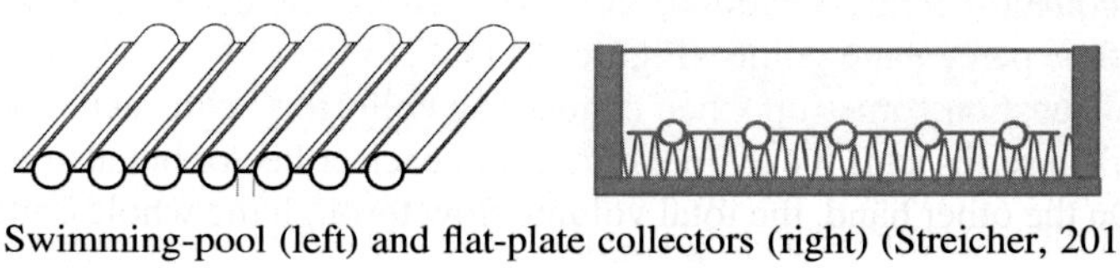

Figure 2.12 Swimming-pool (left) and flat-plate collectors (right) (Streicher, 2015).

reduced. If the total volume flow is chosen with 10 to 15 L/h per m^2 of total collector area, the temperature lift in the whole field is about 40 °C at 1000 W/m^2 irradiance. This is called a low-flow layout according to the low overall volume flow. If the hydraulic layout is around 50 L/h per m^2 of collector area the temperature lift is reduced to 10–15 °C at 1000 W/m^2 irradiance. This layout is called high flow. The hydraulic layout of the whole heating system has to consider the chosen collector flow layout in order not to lose temperature in the system. Losing temperature means that the collector has to operate at higher temperatures and lower efficiencies, respectively, to achieve the same user demand. Figure 2.15 illustrates the different flow layouts for a collector field with four harp-type collectors.

2.3.3 Collector fluid

In climates during which freezing occurs, the collector fluid must stay liquid (or gaseous if air is used). Most often the collector fluid is a mixture of water and 30 to 40% propylene glycol, depending on the lowest ambient temperature at the place of the installation. It has to be considered that the viscosity of that mixture is at low temperatures far higher compared to pure water. This leads to higher pumping pressure loss in the tubes and therefore higher pumping power as well as to lower heat transfer coefficients in the heat exchangers and the solar collector.

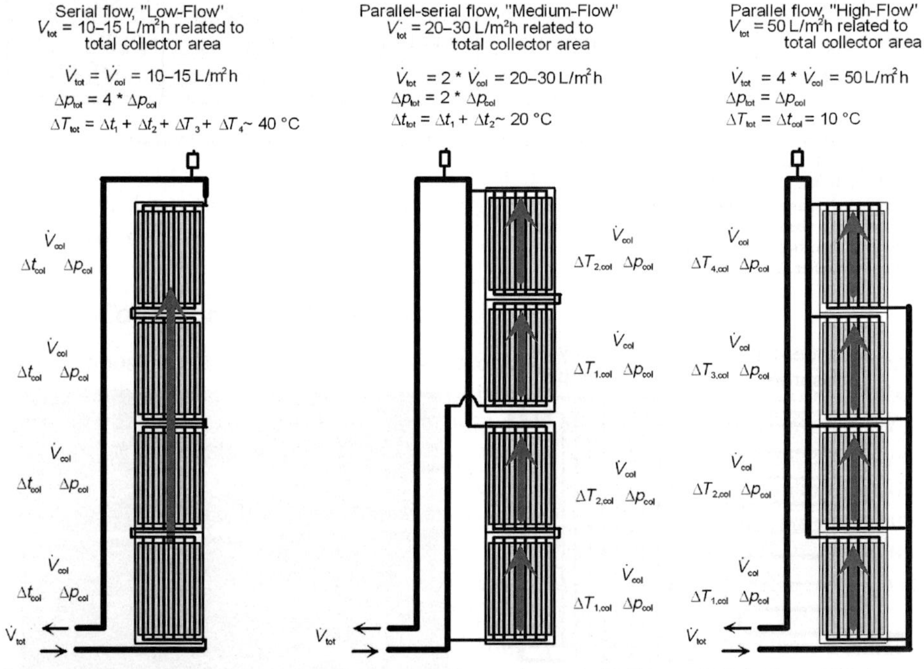

Figure 2.15 Collector hydraulics for low- (left), medium- (middle), and high-flow (right) approach (Streicher, 2015).

Today, mostly premixed collector fluids are used that also include anti-corrosion substances and some alkali reserves.

2.3.4 Stagnation of solar thermal collectors and drain-back systems

When there is no heat needed from the user, all heat storage is already heated up to its maximum allowed temperature and the sun is still shining, the pumps will switch off and the collector goes into so-called stagnation. This situation occurs in solar combisystems every sunny day in summer, as there is no SH demand and the collector field is oversized for the pure DHW demand. Another, but far more seldom possibility, is the breakdown of the electrical grid. This leads to temperatures up to over 200 °C depending on the collector type and the irradiance (see Figure 2.11). Depending on the pressure in the system, the collector fluids start boiling between 100 and 150 °C. Boiling means an increase of the volume by a factor of about 1000.

Figure 2.16 shows two different collector field layouts. In the left layout the connecting tubes to the collector circuit are on the top of the collector field. When the collector fluid starts boiling, the vapor with its great volume is drained into the tubes and down in the basement to the store. The heat exchanger in the heat store is the only place in which the heat can be released and the vapor can be condensed. In this arrangement all collector fluid has to be evaporated. As the propylene glycol is degrading fast at temperatures above 150 °C, this arrangement will make a change of the propylene glycol every few years mandatory. The left layout is, therefore, not recommended.

The layout on the right side of Figure 2.16 has the connecting tubes to the collector circuit at the bottom of the collector field. When boiling starts the vapor generated at the top of the collector (which has the lowest pressure) drains the liquid collector fluid out of the collector. Only little collector fluid has to be evaporated until the whole volume of the collector tubes is filled with steam. Then, only evaporation of remaining droplets and overheating of the steam occurs, but no further liquid collector fluid is drained downward. This system works better the easier the collector fluid can be drained out of the collector. As the initial water is evaporating (fractionating distillation), the propylene glycol never gets to high temperatures, and the lifetime is increased by far. Hydraulic layouts assuring the drainage of the collector fluid by the arising vapor are, therefore, strongly recommended.

Another solution to the problem is the so-called drain-back system. Here, a gas volume is located in the collector loop. As long as the collector pump is running this gas volume is pushed downward into a gas reservoir (drain-back volume) in the loop and the collector liquid can circulate. When the pump stops, the gas volume is moving upward, filling the collector completely, and the water is drained out of the collector. Now only the gas volume is heated up to stagnation temperature. When also all tubes exposed to ambient temperature are filled with the gas volume during stagnation, no antifreeze is needed and the whole collector loop can be driven with pure water. The crucial boundary condition for such a system is that the collector can drain

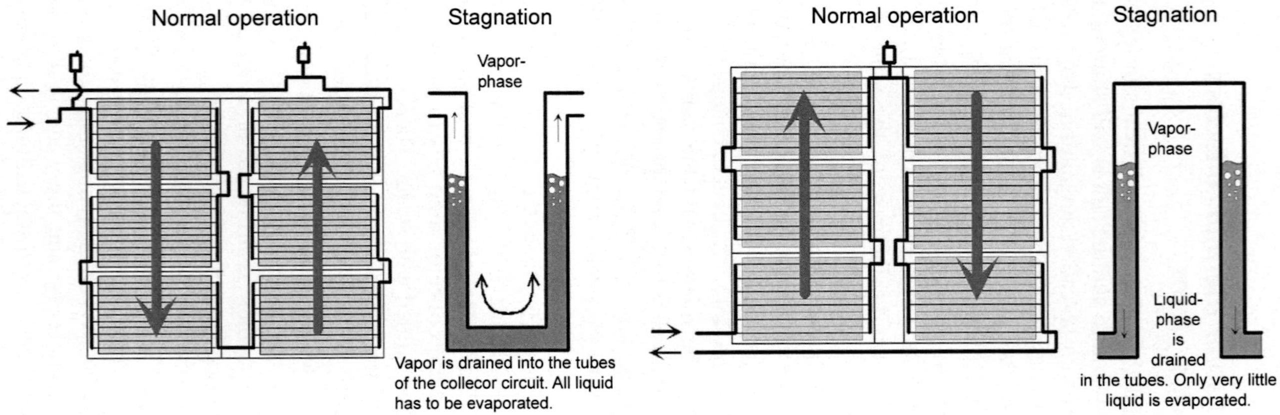

Figure 2.16 Stagnation and boiling in collector fields: unfavorable hydraulic (left), preferable hydraulic (right) (Streicher, 2015).

perfectly, the water can completely flow out of the collector, all tubes are exposed to ambient air during stagnation (no water sack), and the refilling of the circuit when the pump starts works perfectly.

2.3.5 Heat storage

Besides the collector, the heat store is a key element of solar thermal plants. Most often, water is used as storage medium because it has a high specific heat and density ($\rho \cdot c_p = 1.16$ kWh/m^3 K) and a low viscosity. The low viscosity in combination with the dependency on density and temperature (higher temperature is coupled to lower density) allows using the stratification effect in water stores. This means that hot water remains at the top and cold water at the bottom with small horizontal boundary layers between two temperature ranges. If water is heated up at the bottom it is drained upward by buoyancy force and fast mixing with the surrounding water. No "inversion" layer of the temperatures is possible as long as the temperature is above 4 °C. If hot water is supplied in the top of the store, only a little mixing occurs and the cold water below the increasingly hot zone is shifted downward. The same effect takes place when a fully heated store is discharged and cold water is put in at the bottom. In this case the heavier cold water mixes only a little with the hot water above, and the hot "plug" can be nearly completely taken out of the store with constant hot temperature before the cold water "plug" reaches the outlet. If the turbulence in the store can be kept low, the mixing zone can be kept small by heat conduction and convection between temperatures.

These effects are actively used for solar thermal systems. Figure 2.17 shows the different zones of a solar thermal DHW store. The cold water inlet is at the bottom of the store. The heat from the solar collector is also applied at the bottom to keep

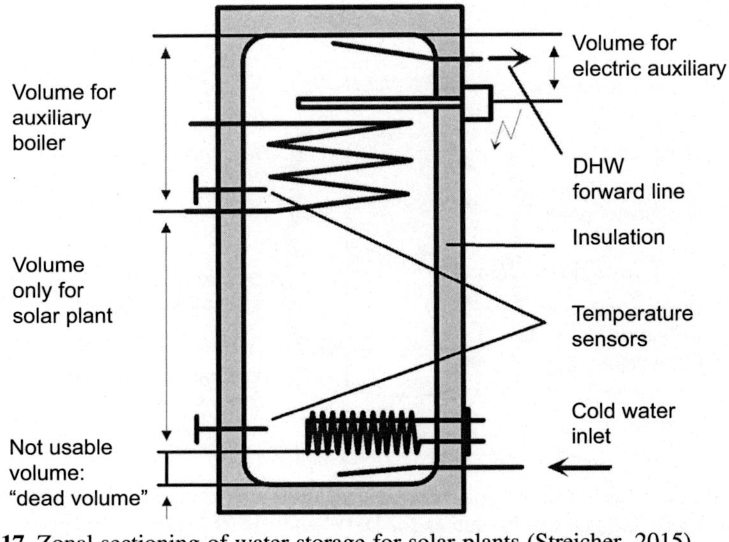

Figure 2.17 Zonal sectioning of water storage for solar plants (Streicher, 2015).

the operating temperature of the collector as low as possible and therefore the efficiency as high as possible. The heat input from the auxiliary heater (boiler and/or electric heater) only keeps the temperature of a volume defined by the maximum user demand of DHW over a period, e.g., 1 h. For single-family houses this volume for the auxiliary boiler is defined by filling a bathtub plus some reserve and amounts to 100 to 150 L. The whole volume between the solar and the auxiliary heat exchanger can be preheated by solar heat. The volume below the solar heat exchanger is called "dead volume" because it is never heated up. Store manufacturers try to build the solar heat exchanger as low as possible.

The maximum setting of the temperature sensor for the solar heat input must be higher than the maximum setting for the auxiliary heater to allow the solar collector to heat also the upper part of the store. When the sun is shining, first the volume of the solar plant is heated until it reaches the temperature of the volume for the auxiliary heater. If the sun is still shining, then the whole volume above the solar heat exchanger is further heated. If the maximum-allowed temperature of solar heating is equal or lower than the set temperature of the auxiliary heater, solar heating can only preheat and the auxiliary burner has to be switched on, e.g., in the night, when the upper part cools down just by heat losses of the store. Additional care by setting the maximum-allowed temperatures has to be given to water hardness (for hard water not to heat above 60 °C) and to lower boundaries due to *Legionella* legislation.

Solar combistores additionally have to satisfy the demand for SH. Most often, the flow/return temperatures for SH are lower than the DHW set temperature but higher than the cold water inlet temperature. Therefore, two more volumes compared to DHW stores can be defined (see Figure 2.18). The volume for SH should be located below the DHW volume but above the preheating volume for the cold water (e.g., if there is a once-through heat exchanger for DHW production or a tank in the tank installation; see Figure 2.20). The volume for the auxiliary heater must cover the DHW volume and part of the SH volume to fulfill both demands, if the sun is not

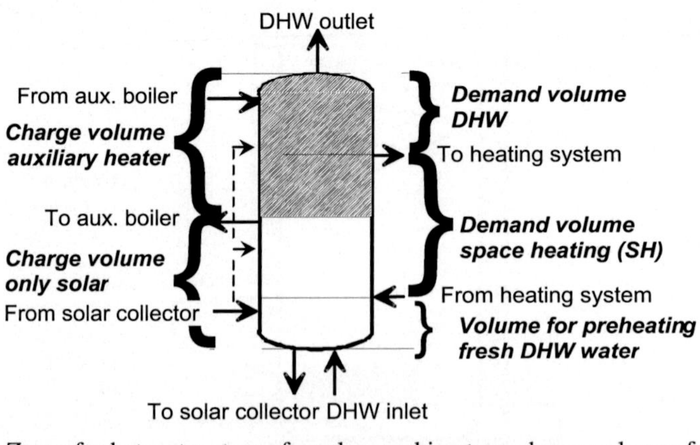

Figure 2.18 Zones for hot water store of a solar combisystem, charge volumes for solar collectors and auxiliary, demand volumes for DHW and SH (Streicher & Bales, 2005).

shining. The charge volume for the solar system remains at the bottom of the tank. As in DHW stores, the maximum-allowed temperatures for the solar part must be higher than those for the auxiliary burner to allow heating of the full store from solar.

There are several ways of transferring heat from the solar circuit to the heat store. If a propylene glycol—water mixture is used for the collector the water in the heat store must be separated from the heat exchanger. Figure 2.19(a) shows the integration via an internal heat exchanger. This arrangement should be coupled to a high-flow collector arrangement, because a high-temperature inlet in the heat exchanger would be lost by immediate mixing of the heated store water when moving upward by buoyancy and the high temperature produced by the solar collector with less efficiency is lost for the user.

The other three solar input schemes have a so-called stratified input. This means that the temperature produced on the store side of the heat exchanger is released in store at the height at which the store temperature matches the inlet temperature. This means that, as little as possible, temperature is lost. Internal stratifiers (Figure 2.19(b) and (d)) are working by updraft of the hotter water in the tube as long as it is hotter than the surrounding water in the store and the outlet to the store when it becomes heavier. Of course, heat loss occurs in the updraft region to the store that reduces the final inlet temperature a bit, but this loss is normally only a few degrees centigrade. Figure 2.19(c) is a so-called external stratifier by valves and two or more inlet positions, which is often used for large solar systems for district heating networks. All three stratifier variants should be used with a low-flow collector layout, as the high temperatures produced in the solar collector at high solar irradiation is put to the top of the store and can be used directly. The solution of Figure 2.19(b) may also be operated with a medium-flow layout. Optimized low-flow systems normally work with slightly better efficiency than high-flow systems. Coupling a low-flow collector layout with a high-flow system significantly reduces the efficiency of the system.

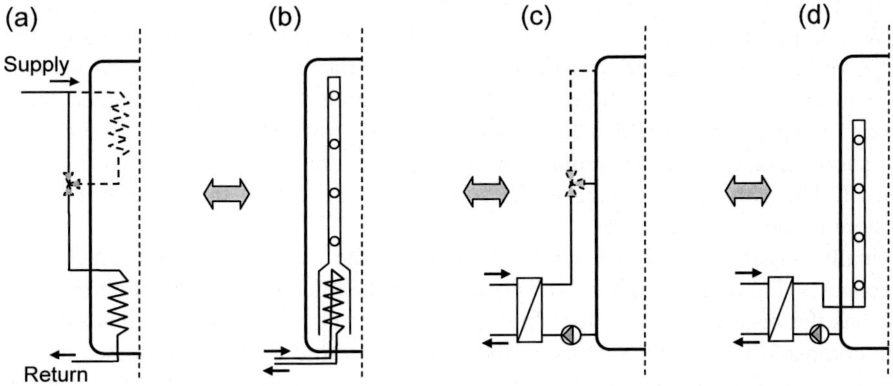

Figure 2.19 Different variants for solar heat input into a solar combistore (heat store with SH and DHW demand). Variants (a) and (c) are both possible with one or more inlet heights each (Haller, 2010). (a) Immersed HX spiral(s). (b) Immersed HX spiral with stratifier. (c) External HX. (d) External HX with stratifier.

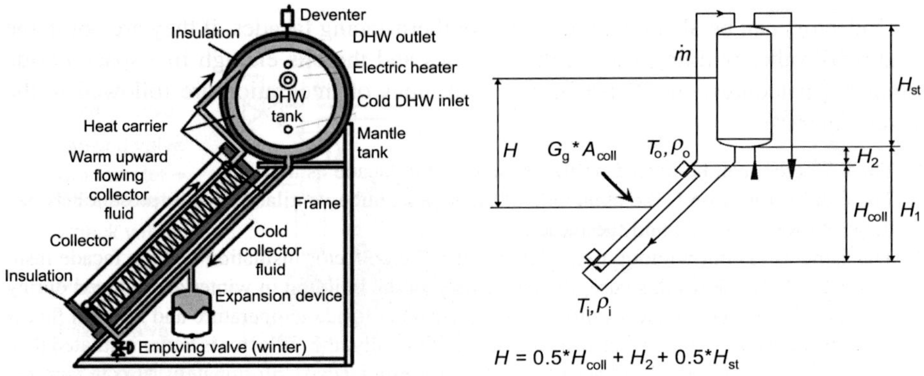

Figure 2.23 Example of a thermosiphon solar DHW system (Streicher, 2015).

Table 2.2 Pipe diameter and insulation thickness for thermosiphon (natural convection) systems EN 12976-2

Collector array aperture area (m^2)	External pipe diameter (mm)	Pipe thickness (mm)	Thickness of one layer insulation[a] (mm)
≥1 and <2	15 ± 1	1	20 ± 2
≥2 and <6	18 ± 1	1.5	30 ± 2
≥6 and <10	22 ± 1	1.5	39 ± 2

[a]Based on a thermal conductivity of 0.04 W/m K ±0.01 for temperature of 10 °C.
Source: EN 12976-2 (2014).

2.4.2 Pumped domestic hot water systems

Pumped systems are needed when the collector is located above the store. This is the case in most Middle and Northern European locations, as the store is not placed in the ambient due to the low temperatures in winter. Figure 2.24 shows the most common hydraulic layout of such a solar-assisted DHW plant. The auxiliary heating may be either electric or by an auxiliary boiler. All layout criteria of store and collector loop will be described in Chapter 3. If a boiler is used, DHW production normally has the priority in SH. Typical dimensions for solar DHW production for single-family houses for Middle European climate are shown in Table 2.3.

2.4.3 Pumped solar combisystems

SH has an inverse demand through the year to the solar radiation. In winter there is high SH demand but low irradiation. In summer there is no SH demand but high irradiation. Therefore, the main use of solar energy for SH is in the periods March to April and September to December. This dependency is shown in

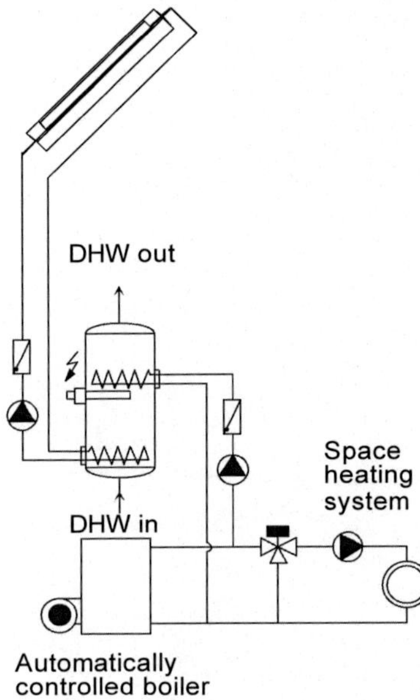

Figure 2.24 Example of pumped solar DHW system (Streicher, 2015).

Table 2.3 Typical dimensions for solar DHW production for single-family houses in Middle European climate

	Collector area (m^2)	DHW store volume (m^3)	Fractional solar saving (%)
2–4 person household	6	0.3	60–80
4–6 person household	8	0.5	60–80

Figure 2.25 for different buildings and collector areas and two climates. The colder the climate and the lower the insulation standard the more the SH demand dominates the total heat demand (SH + DHW).

Various hydraulic systems exist for solar combisystems. Figure 2.26 shows a system with separate stores for DHW and SH production. Such systems are used when a conventional system with SH store already exists (e.g., a biomass log wood boiler) and "upgraded" by a solar thermal system.

New systems are mostly built using synergies of the system and taking into account legislation issues, such as *Legionella* growth protection. They are often built with only one combistore including heat storage functions for DHW and SH. Figure 2.27, left, is a low-flow solar system with stratified solar input (ref. Figure 2.19(d)) and DHW

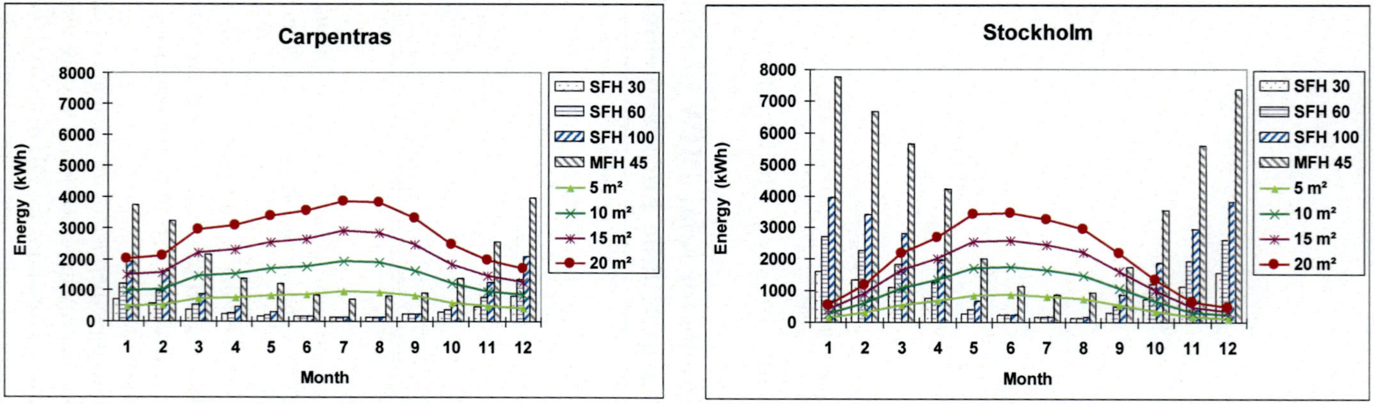

SFH: Single-family house, MFH: multifamily house
Numbers behind SFH and MFH define the energetic quality of the houses and are the space heating demand for Zurich climate in kWh/m²a SH demand.

Figure 2.25 Monthly space heating and domestic hot water demand for four buildings and incident solar energy on different collector areas (facing south, slope of 45°) for Carpentras and Stockholm climates (Streicher and Heimrath, 2003).

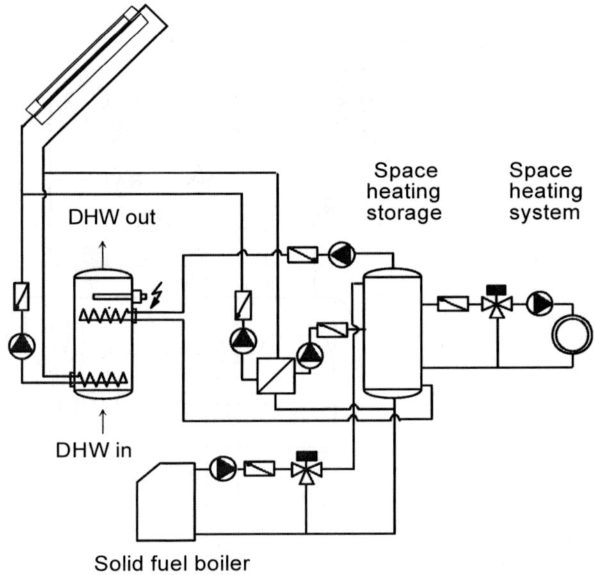

Figure 2.26 Combisystem with separated DHW and SH store (Streicher, 2015). Climates (Streicher and Heimrath, 2003).

production via external heat exchanger (ref. Figure 2.20(c)). Figure 2.27, right, represents a high-flow solar system with two internal heat exchangers. First, the upper part of the store is heated to a temperature high enough for DHW production, and then the lower heat exchanger is activated to make use of the whole store volume. DHW production is done via a tank-in-tank arrangement (ref. Figure 2.20(b)). Both systems follow the rules for the volume zoning of Figure 2.18.

When dimensioning a solar combisystem, of course, climate and building load are the main factors to be considered. The impacts on the demand were shown in Figure 2.25. Other highly important factors are orientation (the azimuth), tilt angle, and size of the collector field, as well as the heat storage volume.

In Figure 2.28 the dependency of azimuth and tilt angle is shown as percentage of the maximum achievable fractional solar saving for a solar combisystem in a multi-family house with 45 kWh/m^2, SH demand, and the climate of Zurich. The highest fractional solar saving of 39% (100% relative) is achieved at an azimuth of 10° west and 55° tilt angle. A vertical arrangement (facade), as well as a horizontal placement to the south, reduces the fractional solar saving by 30%. Variations of the azimuth by ±45° and tilt angle of ±25° from the optimum only reduce the fractional solar saving by 10%. This opens a wide range of roofs for direct integration of solar thermal collectors.

Figure 2.29 shows the dependency of the fractional solar saving on collector size and store volume for all kinds of buildings. The specific collector area (flat-plate collector with selective surface) is calculated by the total collector area divided by the heat load of the building (according to EN 12831, 2003). The specific store volume is

defined as the total store volume of SH and DHW store divided again by the heat load. Several conclusions can be drawn from Figure 2.29:

- For small fractional solar savings small heat stores can also be used, as the DHW demand dominates the whole demand and is constant over the year. Increasing store volume does not give any additional fractional solar savings for specific collector areas of 1 $m^2/k\ W_{heatload}$ as additional solar gain is lost by additional heat loss of the store.
- Even 50% fractional savings can be achieved with relatively small heat store volumes.
- It is very difficult to achieve fractional solar savings above 65%.

2.5 Research and development needs and future trends in technological development, markets, and applications

Future research and development needs are mainly driven by increasing the user acceptance and reducing the costs of the systems to increase the market and be competitive with PV installations. The International Energy Agency published in 2012 a technology roadmap for solar heating and cooling (IEA, 2012). Three goals for decentralized DHW and SH production were given:

- Integrate solar collectors in building surfaces.
- Use alternative materials, technologies, and manufacturing techniques for system cost reduction and performance improvement.
- Address challenges in system design by development of standardized kits and plug-and-function systems.

The first item addresses the issue that solar collectors should become part of the building, reducing the costs by synergetic use of functions for both the building and the solar collector. The second item tries to involve more industrial players on the solar thermal market, whereas the third item intends to reduce the complexity of planning and installation of solar thermal plants, thereby increasing the reliability and assuring the optimal dimensioning and placement of all components of the kit. Several kit systems are available on the market, but they need to reduce their cost and installers have to be convinced to use them, despite the fact that they can charge fewer hours on site during construction.

2.6 Conclusions

Solar thermal collector systems for DHW and SH production have been available on the market for a long time. The worldwide market, driven by China, is steadily increasing. In Europe, the market has decreased since 2008 due to constant investment costs compared to strongly decreasing costs of PV systems. Despite the fact that the technology of solar thermal systems is well known, the development of kit systems already incorporating many components in an optimized way seems necessary. If solar thermal systems for DHW and SH should play a major role in the future, prices have to be significantly reduced.

References

Biermayr, P., Eberl, M., Enigl, M., Fechner, H., Kristöfl, C., Leonhartsberger, K., et al. (2014). *Innovative Energietechnologien in Österreich, Marktentwicklung 2013*. Berichte aus der Energie- und Umweltforschung 26/2104, Bundesministerium für Verkehr. Innovation und Technologie.

EN 12831. (2003). *Heating systems in buildings, method for calculation of the design heat load.*

EN 12975-2. (2006). *Thermal solar systems and components—solar collectors—Part 2: Test methods.*

EN 12976-2. (2014). *Thermal solar systems and components—factory made systems Part 2: Test methods.*

Haller, M. (2010). *Combined solar and pellet heating systems-improvement of energy efficiency by advanced heat storage techniques, hydraulics, and control* (Ph.D. thesis). Graz University of Technology, Institute of Thermal Engineering.

Heimrath, R. (2003). *Generic system #19: Centralized heat production, distributed heat load.* Technical Report for Subtask C, IEA SHC-Task 26, Appendix 9 to [5].

IEA. (2012). *Technology roadmap solar heating and cooling*. Paris, France: International Energy Agency. www.iea.org.

Kaltschmitt, M., Streicher, W., Wiese, A., & (Hrsg). (2013). *Erneuerbare Energien, Systemtechnik, Wirtschaftlichkeit, Umweltaspekte* (5. Auflage). Springer Verlag, ISBN 978-3-642-03248-6. ISBN:978-3-642-03249-3 (eBook).

Mauthner, F., & Weiss, W. (2014). *Solar heat worldwide, markets and contribution to energy supply 2012, IEA Solar Heating & Cooling Programme, June 2014.*

SPF. (2015). *Institut, für Solartechnik.* Rapperswil, Switzerland http://www.solarenergy.ch/Kollektoren.111.0.html. Accessed 01.03.15.

Streicher, W. (2015). *Solar thermal energy use, lecture book.* University of Innsbruck.

Streicher, W., & Bales, C. (2005). Combistores. In J. C. Hadorn (Ed.), *Thermal energy storage for solar and low energy buildings—State of the art by the IEA solar heating and cooling task 32.*

Streicher, W., & Heimrath, R. (2003). *Analysis of system reports of task 26 for sensitivity of parameters.* A Report of IEA SHC-Task 26, Solar Combisystems December 2003. TISUN/DTEC 2011, Söll, Austria.

Solar thermal process heat (SPH) generation

3

Stefan Hess
Stellenbosch University, Stellenbosch, South Africa

3.1 Introduction

Industrial process heat accounts for more than two-thirds of the total global industrial energy consumption. Half of this demand occurs at temperature levels below 400 °C. Approximately 40% of the current industrial energy consumption is covered by natural gas and approximately 40% by petroleum products (International Renewable Energy Agency (IRENA), 2015). In several industry sectors such as food, wine and beverages, textiles, transport equipment, machinery, or pulp and paper, the share of heat demand below 250 °C is around 60% (Potential of Solar Heat for Industrial Processes (POSHIP), 2001). There is a wide variety of promising applications in industry, commercial services, and agriculture. Common is, e.g., the preheating of water for washing or cleaning and the preheating of boiler makeup water. Solar heat can also be integrated directly into industrial heat supply networks, either by heating pressurized water or by solar steam generation. Furthermore, solar air collectors can be used to preheat air for drying or combustion, as well as to heat production halls.

3.2 Potential

3.2.1 Germany

To illustrate the process heat demand and technical potential for solar thermal process heat (SPH) in a highly industrialized country, the example of Germany is used. Figure 3.1 shows that the industrial sector accounts for 28.9% of Germany's final energy demand. Almost three-quarters of its demand is heat. The process heat demand in the service sector is lower, but the temperature levels are often very favorable for SPH. Together, process heating in the German industry and service sector accounted for 19.8% or 494 TWh of totally 2499 TWh consumed in 2012.

Figure 3.2 shows a summary of the temperature distribution when the processes of the 11 most relevant German industry sectors are considered. About two-thirds of Germany's industrial heat demand must be provided above 500 °C. A share of 29% of the industrial heat demand is required at favorable temperature levels for SPH below 200 °C. The 21% below 100 °C include also hot water demand (HW, 1%) and space heating (SH, 11%), so that the share of heat demand in the intervals below 100 °C and

Renewable Heating and Cooling. http://dx.doi.org/10.1016/B978-1-78242-213-6.00003-5

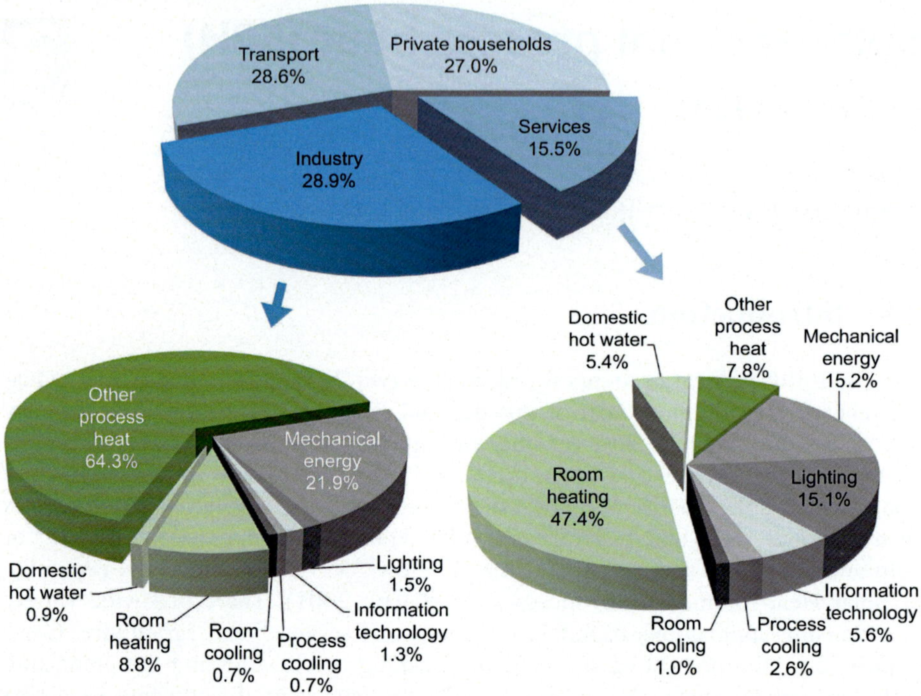

Figure 3.1 Distribution of Germany's final energy consumption in 2012 (top) with a closer look into the sectors of industry (left) and services (right).
Source: Hess (2014).

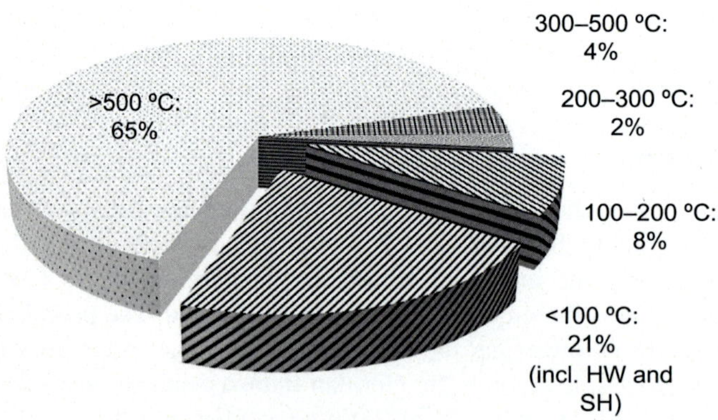

Figure 3.2 German industrial heat demand within different temperature ranges.
Based upon values of Lauterbach et al. (2012).

between 100 °C and 200 °C is about the same. It is remarkable that the demand in the range from 200 °C to 500 °C is comparably low.

Based on these data, Lauterbach et al. (2012) assessed the SPH potential for Germany. They considered only the most relevant 11 industrial sectors and included only processes supplied below 300 °C. Further, they assumed that 60% of the theoretical potential could not be realized due to the priority of energy efficiency measures, the necessity of electrical heat supply (e.g., for plastic products), or limited area to install the collectors. For the remaining demand they suggest an average solar fraction of 30%. These restrictions result in a technical potential of approximately 3.3% of Germany's overall industrial heat demand (468.9 TWh/a), corresponding to 16 TWh/a. This is approximately 11% of the industrial heat demand below 300 °C. A detailed description of the 11 industrial sectors with assessments of their suitability for SPH can be found in Lauterbach et al. (2012).

3.2.2 European Union

Industry accounts for about 44% of the EU's overall heat demand, with about two-thirds of the heat required above 250 °C and one-third below (European Technology Platform on Renewable Heating & Cooling (RHC-Platform), 2011). In the framework of the International Energy Agency (IEA) Solar Heating and Cooling (SHC) Task 33/IV, Vannoni et al. (2008) extrapolated the potentials found by older studies on five European countries. They roughly estimated that a share of 3.8% of the industrial heat demand within the EU 25 could be generated by state-of-the-art solar thermal (ST) systems. Nonindustrial process heat demand was not considered in this study. The potential given by Vannoni et al. corresponds to about 190 TWh/a, or approximately 423 million m^2 (assuming solar gains of 450 kWh/m^2a). The magnitude of this potential can be assessed as follows: By the end of 2012, a cumulated collector area of 61 million m^2 was installed in the EU 28 for all ST applications (calculated according to Weiss and Mauthner (2014, p. 8) from 42.8 GW_{th} installed with a factor of 700 W/m^2).

3.2.3 Worldwide

IRENA did a global assessment of process heating requirements. In a best-case scenario with considerable cost reductions for ST and without fossil fuel subsidies in industry, it estimated a technical potential for SPH of 15 EJ or 850 GW_{th} (out of 160 EJ assumed process energy demand in 2030). This high technical potential is reported mainly caused by the necessity of new capacity investments, e.g., in the chemical sector. About 80% of the capacity to install would be located in non-OECD countries (IRENA, 2015).

The economically realizable potential is significantly lower, but still high in absolute figures. Taking into account fossil fuel prices and technology learning curves for solar process heat, IRENA estimates that a share of 3.3 EJ (c. 180 GW_{th}) could be realized (IRENA, 2015). For 2050, the United Nations Industrial Development Organization (UNIDO) estimates 5.6 EJ (c. 300 GW_{th}) of SPH (UNIDO, 2011). In addition, the IEA stresses the high potential for low-temperature SPH. It expects that ST systems globally could supply 20% of the industrial heat demand below 120 °C until 2050 (IEA, 2012).

Details on the shares of heat demand at different temperature ranges and short descriptions of the processes in every sector can be found in Lauterbach et al. (2012). Worldwide, also mining processes, drying processes, and solar (community) cooking are very promising applications.

"Low-hanging fruits" for SPH integration are processes in which

- The consumption of expensive energy carriers can be reduced by SPH (most favorably electricity, but also diesel, heating oil, paraffin, liquefied petroleum gas (LPG), etc.).
- A constant heat demand occurs during a minimum three-quarters of the year, including summer, as well as at least 5 days per week.
- The process return temperature is below 50 °C (beneficial for stationary collectors).
- Heat recovery from other processes is technically or economically not feasible.
- Potential future changes are unlikely to affect the efficiency of an SPH installation.

3.5 Solar thermal technologies for process heating

Size and temperature level of SPH projects are often in between domestic and CSP solar installations. A high variety of adapted collector, storage, and system concepts has been developed for this market. For preheating and low-temperature processes below 90 °C standard domestic systems are applied. These can be low-pressurized or non-pressurized and with forced circulation (pump) or natural circulation (thermosiphon). Temperatures between 90 and 150 °C can be provided by pressurized systems with improved nontracking collectors. These systems usually have pressurized water storage. Usually water is used as a heat carrier; in regions with the necessity of freeze protection, glycol is added. Also small, one-axis tracked process heat collectors like small Fresnel collectors or parabolic troughs with and without storage are applied. Above 150 °C, usually tracking collectors with the heat carriers pressurized water, thermal oil, air, or molten salts are applied. Some technologies are able to directly generate steam within the receiver. Above 400 °C, exclusively two-axis tracked collectors like parabolic dishes or central receiver heliostat fields with molten salt as heat carrier are used.

3.5.1 Process heat collectors

Process heat applications demand heat from very low to very high temperature levels, so almost every ST collector type can be used for process heat generation. Table 3.1 gives an overview of common process heat collector technologies.

ST collectors are either of the nonconcentrating or concentrating type. Nonconcentrating collectors like FPCs or standard ETCs do not increase the natural flux density of the solar irradiance until it is absorbed by their absorbers. By concentration, the area for radiation collection (aperture area) is decoupled from the area converting the radiation into heat (receiving absorber, in which thermal losses occur). Reflectors re-direct solar irradiance toward the receiver, so that the radiant flux density on the receiver is higher than on the aperture. This concentration can either be nonfocusing or focusing.

Table 3.1 Common collector types for SPH generation with indicative working temperatures (collector symbols from Norton, 1992)

Tracking	Collector type	Symbol	Absorber type	Concentration	Indicative working temperature range (°C)
Stationary (none)	Standard flat-plate collector (FPC)	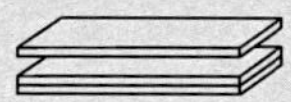	Flat	No	30–90
	Standard evacuated-tube collector (ETC)	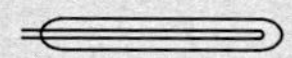	Tubular	No	50–130
	Improved stationary collectors with and without reflectors		Tubular/flat	Some yes, some no	80–150
Single axis	Linear Fresnel collector (LFR)		Tubular	Yes	60–400
	Parabolic trough collector (PTC)	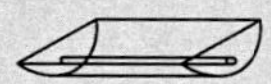	Tubular	Yes	100–450
Two axes	Parabolic dish collector (PDC)		Point	Yes	100–500
	Heliostats with central receiver (HCR)		Point	Yes	150–2000

Nonfocusing concentrators like the compound parabolic concentrator (CPC) directly reflect irradiance from a certain range of incidence angles (acceptance angle) onto a receiver. This way, low concentration ratios (aperture area divided by receiver area) can be achieved without tracking and a share of the diffuse irradiance is used. Focusing concentrators direct the solar beam irradiance to a point or line receiver. Common line-focusing collector types are parabolic trough collector (PTC) and linear Fresnel collector (LFR); they track the sun in one axis. Maximum temperatures can be achieved by point-focusing collectors like parabolic dish collector (PDC) and Heliostats with central receiver (HCR), which track the sun in two axes.

The characteristic performance parameters of ST collectors are their thermal efficiency curve and their incidence angle modifier curves (within the collector axes not tracked). Test results on performance and quality of the collectors certified according to the European Solar Keymark Label can be found on the Solar Keymark website (Solar Keymark). Also the results for products tested according to the American Solar Rating and Certification Corporation are available online (SRCC).

3.5.1.1 Air collectors

Low-temperature collectors with air as heat transfer fluid can be flat plates or evacuated tubes. They are used for drying of food products, wood or wood chips, but also for preheating of combustion air for boilers. Direct solar dryers are filled with the product and have a transparent cover, so that sun rays directly reach the product. In case of food products, the UV irradiance can reduce the quality of the product. Usually, air collectors are used in indirect drying processes, so that the product is not directly exposed to solar irradiance. Figure 3.4 shows a solar air collector with rear-flow absorber. These collectors can be linked serially to large rows, with special collector models for the first and last collector of each row. Other air collector types have a front-flow absorber (air between absorber and cover) or volumetric absorbers for increased heat conduction to the air flowing through.

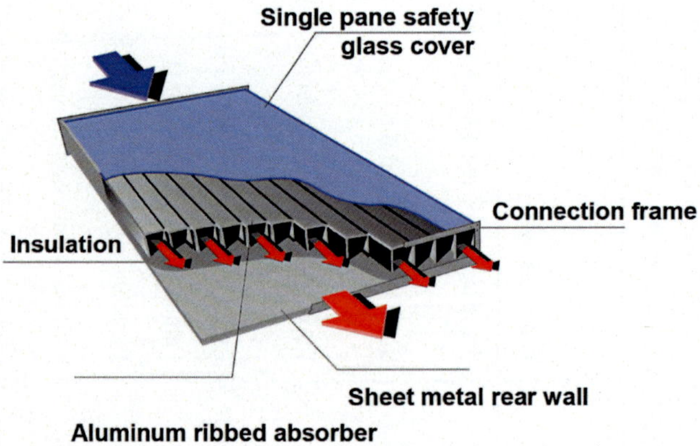

Figure 3.4 Solar air collector with rear-flow absorber.
Source: Grammer Solar.

3.5.1.2 Improved stationary collectors

Standard flat-plate and evacuated-tubular collectors are described in detail in Chapter 2 of this book. They are very mature technologies, well suited for low-temperature SPH generation. Major loss mechanisms of nontracking ST collectors are conduction, convection, and thermal radiation. To efficiently generate temperatures between 80 and 150 °C, one or several of these loss mechanisms have to be reduced. Recent stationary collector developments of this kind include:

- Double antireflective glazing for flat plates (optionally with a low emissivity coating on the outward-pointing surface of the inner glass pane)
- Plastic foil as a second transparent cover between antireflective glass pane and absorber
- Transparent insulation material as collector cover
- CPC reflectors for flat plates or evacuated-tube collectors, either only redirecting, irradiance, or concentrating solar irradiance at collectors with wider tube distance (concentration ratios up to 4 are achieved, but ratios around 1.5 are typical)
- Use of inert gases or vacuum in flat-plate collectors

The standard heat transfer fluid below 150 °C is water, with an addition of glycol for freeze protection if necessary. Rarely, also air and steam can be found.

Figure 3.5 shows the RefleC collector as an example for an improved stationary flat-plate collector. At this concept, booster reflectors fill the gap between flat-plate collector rows, which increase solar gains in summer without additional roof or ground area necessary. The collector has been installed in the laundry Laguna in Germany, where glass-foil double covered flat-plates with and without booster reflector are compared in a collector field of 65 m^2 aperture at working temperatures up to 135 °C. The field heats washing water, boiler makeup water, and feed water. At this plant, including all collector inlet temperatures above 80 °C, an additional annual energy gain of 78% caused by the booster reflectors was measured (Hess, 2014).

Figure 3.5 Low-concentrating stationary flat-plate collector with transparent double cover (anti-reflective coated (AR)-glass and ethylene/tetrafluoroethylene copolymer (ETFE)-foil) and booster reflector approximating a one-sided CPC (Hess, 2014).

An advantage of stationary, nonfocusing concentrators is their ability to also use a share of the diffuse irradiance. A drawback is the typically low concentration ratios and thus the much larger absorber areas compared to focusing collectors. Increasing heat losses due to thermal radiation at higher working temperatures limits the application temperature range of stationary concentrating collectors.

3.5.1.3 Parabolic trough collectors

Figure 3.6 shows a PTC. Its reflector has a parabolic cross-section. PTCs are line-focusing collector types; i.e., the absorber tube is positioned in the focal line of the trough. The theoretical plane extending between the vertex of the parabola and the receiver tube is called the longitudinal plane. All light reaching the reflector parallel to the longitudinal plane is directed toward the absorber. Thus, PTCs have to track the sun to maintain this condition and they can only use beam irradiance.

The PTC trough shown in Figure 3.6 belongs to a field of 627 m^2 aperture area, which is installed at the cheese factory of Emmi AG in Saignelégier, in the Swiss Jura Mountains. Solar heat is integrated into the hot water network of the plant, generating about 50% of the heat demand of the plant at bright summer days and saving 30,000 L of heating oil per year.

As reflective material, usually a very thin silver layer below white glass is used, but also concepts using outdoor-stable, anodized aluminum reflectors exist. As receivers, usually evacuated glass tubes with selectively coated-steel absorber pipes are used. Also nonevacuated receiver tube concepts exist. Here, the glass tube reduces conduction and convection losses. There are also concepts in which a flat glass pane extends from one edge of the reflector to the other, protecting the trough against environmental impacts and reducing thermal losses. This has to be paid by increased optical losses. That's also the case for concepts in which a whole collector field is enclosed by a greenhouse, protecting reflectors and tracking mechanisms from

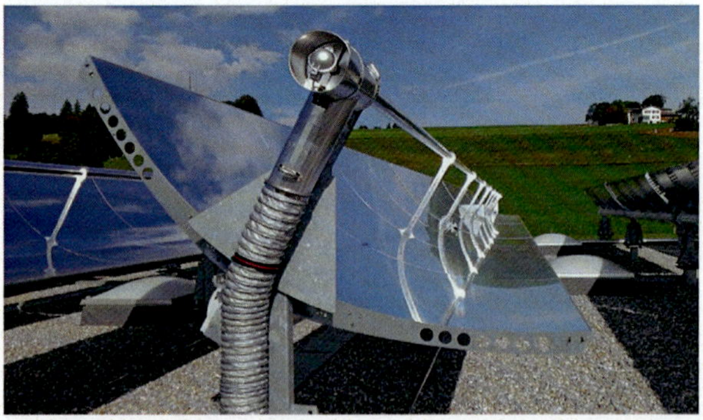

Figure 3.6 Small parabolic through collector with nonevacuated receiver tube and aperture width of 1.85 m.
Source: NEP Solar.

sandstorms in desert environments. Such concepts usually require regular cleaning of the cover glazing.

The optimal orientation of the tracked collector rows and the necessary distance between the rows depend on several factors, such as latitude, trough length, heat demand profile, and temperature level of supported processes. Near the equator, usually a North–South orientation of the troughs is beneficial. In this case, the incident irradiance has low longitudinal incidence angles, resulting in high conversion efficiencies and low so-called end losses during the course of the day. End losses mean beam irradiance reflected, but spilling over the trough in the longitudinal direction, not hitting the absorber tube. East–West orientation is optimal for high latitudes. It requires less tracking effort than North–South orientation, but in the morning and the afternoon end losses have to be accepted.

Below 150 °C usually pressurized water is applied, with glycol for freeze protection if necessary. Between 150 and 200 °C pressurized water or thermal oil are applied. From 200 to 400 °C usually thermal oils are used. Also direct solar steam generation within the collectors is possible.

3.5.1.4 Linear concentrating Fresnel collectors

Figure 3.7 shows a Fresnel collector. The fundamental difference to PTCs is that the receiver tube is stationary and the primary reflectors are individually tracked to focus beam irradiance toward the receiver.

The receiver of LFRs is usually covered by a CPC-shaped secondary reflector. It directs the beam irradiance reflected by the primary reflectors toward the absorber and also protects the receiver from environmental impacts. When nonevacuated receiver tubes are applied, the secondary reflector casing is often thermally insulated

Figure 3.7 Fresnel collector module with an aperture width of 7.5 m, evacuated fixed receiver tube, and secondary concentrator.
Source: Industrial Solar.

and closed by a glass pane at the lower side to reduce thermal losses. Application temperatures and working fluids of LFRs are similar to those of PTCs.

LFRs have lower wind loads than PTCs and require only a small distance between the rows for mounting and maintenance. This predestines them for factory roof mounting. PTCs on the other hand can to a certain extent compensate for low solar altitudes, because their aperture area is tracked. At high latitudes (i.e., low solar altitudes), LFRs usually have higher optical losses than PTCs, because their aperture plane remains horizontal, which results in higher cosine losses as well as blocking and shading between the primary reflectors. Because the distance between primary reflectors and receiver is usually much higher than the focal width of a PTC, LFRs require higher module lengths to operate efficiently. For both PTCs and LFRs, comparably small collector concepts with aperture widths below 1 m also exist. These collectors are sloped toward the sun and can be used, e.g., for façade integration.

The two LFR strings shown in Figure 3.7 each consist of 11 modules. The field has a gross aperture area of 484 m^2, operates with pressurized water at 180 °C and 16 bar, and has a thermal peak power of 242 kW. It drives a double-effect absorption chiller producing cold water for the company MTN in Johannesburg, South Africa. An uninterrupted power supply unit is installed to drive the mirrors into protective stow position, if power to the system is cut. In case of rain, the mirrors automatically move into a self-cleaning position.

3.5.1.5 Parabolic dish collectors

The reflector of a PDC has the shape of a paraboloid, so that beam irradiance is focused onto one focal point. To maintain this focus, the dish has to track the sun in two directions throughout the day. Double-axis tracking involves usually higher accuracy but also mechanical effort, e.g., to resist wind forces. But point-focusing systems have significantly lower heat losses than line-focusing ones due to the higher concentration ratio.

Figure 3.8 shows a large dish, in which the parabola is approximated by small, fixed-slope Fresnel segments positioned in the aperture area. The system generates steam at Ramakrishna Mission Students Home in Chennai, India. On a sunny day, up to 600 kg of steam is produced, suitable to cook about 4000 meals per day. Because the system uses pressurized hot water storage, in the morning steam for cooking can also be made available by reducing the storage pressure.

3.5.1.6 Heliostats with central receiver

Heliostats offer the highest concentration ratios of all ST collectors. Flat, double axis tracked reflectors (heliostats) focus beam irradiance onto a stationary, central receiver tower with flat or cylindrical absorber. Usually, each heliostat is individually controlled. Besides steam or thermal oil, below 500 °C also molten salt is used to remove the heat from the receiver. Above 500 °C, receivers with air as heat transfer fluid can be used. To reduce the convective heat losses, some designs have the absorbing surface placed inside a cavity receiver.

Figure 3.8 ARUN-100 dish, a two-axis tracked focusing collector with an aperture area of 100 m^2 and pressurized water storage in front.
Source: Clique Solar.

Figure 3.9 shows a heliostat field with a flat receiver mounted on a tower of 100 m height. The plant has a thermal power of 29 MW and is located in Coalinga, California, USA. To increase the effective reflector area (i.e., to reduce cosine losses), the tower is located at the southern edge of the field. The plant generates steam for enhanced oil recovery (EOR). With this technique, steam is pressed into the ground to solve heavy

Figure 3.9 Solar power tower plant with 3822 heliostats, each consisting of two reflectors with c. 25.5 m^2, resulting in an aperture area of 194,000 m^2.
Source: Brightsource Energy.

oil, increasing the output of existing oil pumps. The solar steam replaces natural gas, which was previously used as the main heating source for EOR.

3.5.2 Storage

Storage of SPH is very important if the process heat demand is not continuous or if high solar shares of the heat demand shall be covered by solar heat. Small and medium enterprises often work only 5 days per week. In this case it is usually economic to store the solar gain from Saturday and Sunday to use it during work.

Generally, existing storages should be used whenever technically possible. In breweries, for example, often the brewing water storage can be heated. To form an effective solar buffer storage tank, only the addition of an insulating layer is necessary. At certain processes, the use of process internal storage is also possible. This is, for example, the case in surface-treatment factories if the bath temperatures can vary.

Generally, the available storage technologies can be categorized into sensible, latent, and chemical storage. The majority of SPH plants installed worldwide are used for applications below 100 °C. Thus, currently the standard solution for SPH storage is sensible heat storage, mostly using water buffer tanks. Above 100 °C, pressurized water storages with improved insulation, e.g., by vacuum, are available. Because the storage costs rapidly increase with pressure, with increasing temperatures thermal oil storage and also molten salt can be considered. For high temperatures, concrete block storage or rock bed storage (for air heat carrier) is an option.

Latent heat storage has recently entered the market of domestic solar buffer storage tanks. Phase Change Materials can be placed within conventional water storages to increase the storage capacity per volume and to reduce heat losses. This is possible because latent heat is stored at a certain, constant temperature level, depending on the material used. These storages may also be applied in SPH plants in the future. Many promising concepts for latent heat storage above 120 °C exist (cp. Tamme et al., 2008) but have only been applied in lab-scale projects so far.

In chemical storage concepts, a product is separated into its reactants by the addition of heat. The heat can be released again when the separated components recombine into the product in an exothermal reaction. Theoretically, this storage type is very promising, because no heat losses occur over time. However, the technical problems with heat conduction and product separation are still tackled in research projects and no commercial product is yet on the market.

For some applications it is possible and favorable to size the SPH system in a way that no storage is needed. In this case, the process should run whenever solar heat is produced and the solar peak power should be below the minimum required power of the industrial plant. This is applies, e.g., to direct solar steam generation parallel to an industrial boiler in companies with high energy demand. Such companies often work three shifts per day at 7 days a week to avoid cooling down the steam network. In many of these cases, covering even the entire roof with steam-generating collectors might only supply a single-digit solar fraction of the annual steam demand. Such SPH systems can be very cost effective and easy to integrate into the existing heat supply.

3.5.3 Integration of solar heat

The integration of solar heat into industrial, commercial, or agricultural plants is usually quite complex. Prior to planning of an SPH plant, energy efficiency measures (heat recovery, insulation of pipes, etc.) and process optimization (reduction of the energy demand per unit of product, e.g., by new machinery) have to be checked. These measures usually have significantly shorter amortization periods than SPH. Additionally, overdimensioning of SPH systems can only be avoided if potential future reductions of the heat demand can be realistically assessed.

Solar heat can be integrated at the supply and/or process level of an industrial plant. On the process level, solar heat is supplied to selected thermal processes. On the supply level, solar heat is supplied to the heat distribution network of the plant (steam or hot water network). This way, all thermal processes connected to this network are indirectly supported. Here, the required temperature level is usually higher, which decreases the solar efficiency. On the other hand, the future performance of the SPH plant is much less affected by changes of single processes.

Low-temperature SPH systems often consist of five different parts, which are illustrated in Figure 3.10. Solar heat is generated in the collector loop and charged to buffer storage whenever the temperature at the bottom of the storage is lower than the temperature at the collector outlet. When there is process heat demand and the temperature at the top of the buffer is higher than the process return temperature, the storage is discharged and SPH is integrated into a certain process.

Preheating systems can usually be realized with stationary collectors and at low costs. The decisive temperature level in this case is the process return temperature, because it determines the storage temperature at its bottom and thus the collector inlet temperature during operation. The lower this temperature, the better. Figure 3.11 shows a solar heating system.

In this example, the SPH system heats a galvanic bath, which has to be kept at a certain temperature level. The fundamental difference to preheating systems is that the integration return temperature is higher than the process temperature, which reduces the efficiency of the SPH system. To use the buffer storage only when required,

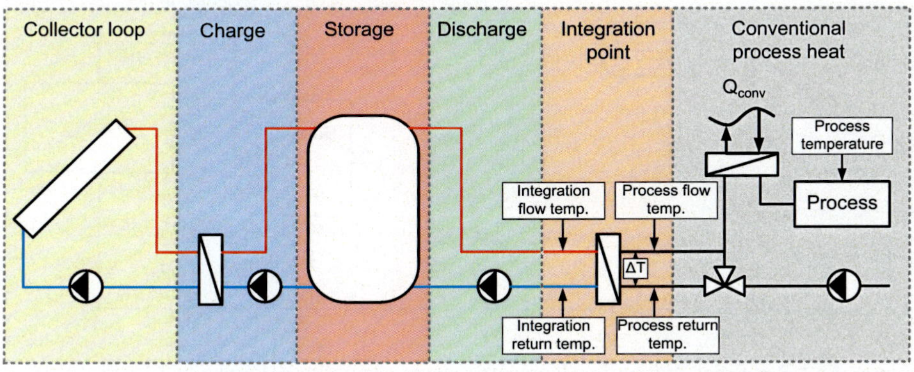

Figure 3.10 Simplified solar thermal system for preheating (Helmke and Hess, 2015).

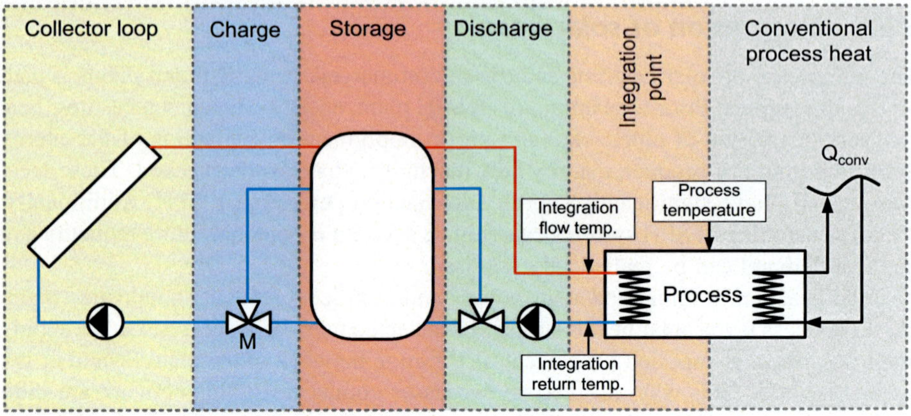

Figure 3.11 Simplified solar thermal system for heating on process level (Helmke and Hess, 2015).

the return flow to the storage is stratified. Additionally, a mixing valve on the charging side assures that only the top part of the storage is charged when the process is active and no mixing of the storage occurs.

Figure 3.12 shows the principle of direct solar steam generation. The focusing collector acts as a parallel steam boiler, which makes integration easy. The conventional steam boiler has to be able to modulate by the extent the solar loop maximally contributes to the steam demand.

When the beam irradiance exceeds a certain level and there is steam demand, boiler feed water is circulated through the collector loop by the collector pump until it starts evaporating. The steam is separated from the fluid in a steam drum and released to the industrial steam line, and additional feed water enters the collector loop. Recent system developments aimed at linking the collector loop directly to the boiler, using the boiler as a steam drum.

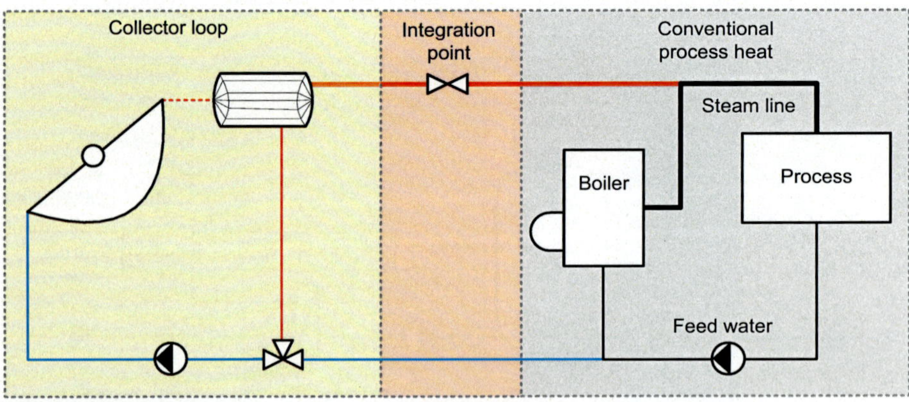

Figure 3.12 Simplified system for direct solar steam generation (Helmke and Hess, 2015).

A detailed methodology to successfully integrate solar heat into industrial processes and on the different phases of realization of an SPH project is described in Muster (2015).

Examples for SPH system design to find a good interrelation between solar fraction and specific solar yield are discussed in Hess and Oliva (2010) and by Hess et al. (2011).

3.6 Examples

In the following, four low-temperature applications for SPH are described. Examples with focusing collectors are reported in Häberle (2012), SHIP-plants (2015), and CSH India (2015).

3.6.1 Brewing

Breweries have a high low-temperature demand and often a high potential for process optimization. Manufacturing of malt and beer requires large amounts of electrical and thermal energy. According to Mauthner et al. (2014), the entire heat demand of breweries and malting plants can be supplied at temperature levels between 25 °C and 105 °C, with processes like malt drying, bottle cleaning, pasteurization, or mashing being very promising for SPH.

Sector-specific SPH concepts for this sector have been developed, and a number of successful SPH projects have been carried out recently. Figure 3.13 shows a flat-plate collector field at the Goess Brewery in Leoben, Austria, which was installed in June 2013.

Two steam-supplied vessels (mash tuns) were retrofitted with internal plate heat exchangers to allow the mash tuns to be heated with hot water instead of steam. This hot water supply is fed by waste district heat from a biomass plant as well as from the

Figure 3.13 Flat-plate collector field of 1375 m^2 at the Goess Brewery.
Source: AEE INTEC.

Figure 3.16 Field of 80 high-performance evacuated-tube collectors with CPC reflectors on the roof of Zehnder in Gränichen, Switzerland, with a gross area of 394 m^2.
Source: Ritter XL Solar.

The radiators are pretreaded in baths at a temperature of c. 60 °C before they are coated. About 40% of the heating demand of Zehnder is caused by these baths. Figure 3.16 shows the solar collector field of the company, which was commissioned in July 2012. The field is oriented along the roof edge with about 25° turned toward the southeast. The collectors are sloped 45° and unshaded. Because the production hall was newly constructed before the collectors were installed, the necessary substructure for the collectors had already been considered beforehand.

Figure 3.17 shows the integration of the plant into the hydraulic system. Solar heat is directly integrated into the heating network of the new production hall; so the collector field works like an additional boiler. The installation is a so-called Aqua System, operating with water only as a heat carrier fluid. Thus, no heat exchanger between collector loop and heating network is necessary. Depending on the current power of the field, the solar loop is either used as return flow boost for the boilers or to deliver the set flow temperature of 85 °C to heat the baths.

Because the bath itself can be used as a buffer, only a very small storage tank of 5 m^3 is installed. This storage acts as a hydraulic separator, compensating the different mass flows through the solar loop, the boilers, and the baths.

Ritter XL Solar states that at very good summer days the plant covers the whole heating demand of the baths. Annually, the solar fraction at this process is 30%. Alternatively, the heating of the production hall can also be supplied by solar.

Considering all material and installation costs of the system including the steel sub-structure for mounting, the solar plant had specific investment costs of about 1000 EUR/m^2, not including subsidies. Thanks to a grant from the Swiss Federal Office of Energy, the plant is expected to have a payback period slightly above 5 years.

According to Ritter XL, the simulated annual gain of the system is 400 kWh/m^2a, summing up to 158 MWh/a, which means annual savings in LPG of about 200 MWh. The system is a turnkey system and was installed under Guaranteed Solar Results Contracting. In this case, this means if the system delivers less than 90% of the simulated annual energy gain as a mean value over the first 3 years, the solar company is paying the costs of the fuel not saved. In the first 2 years of operation, the measured solar gains matched the simulated system gains very well.

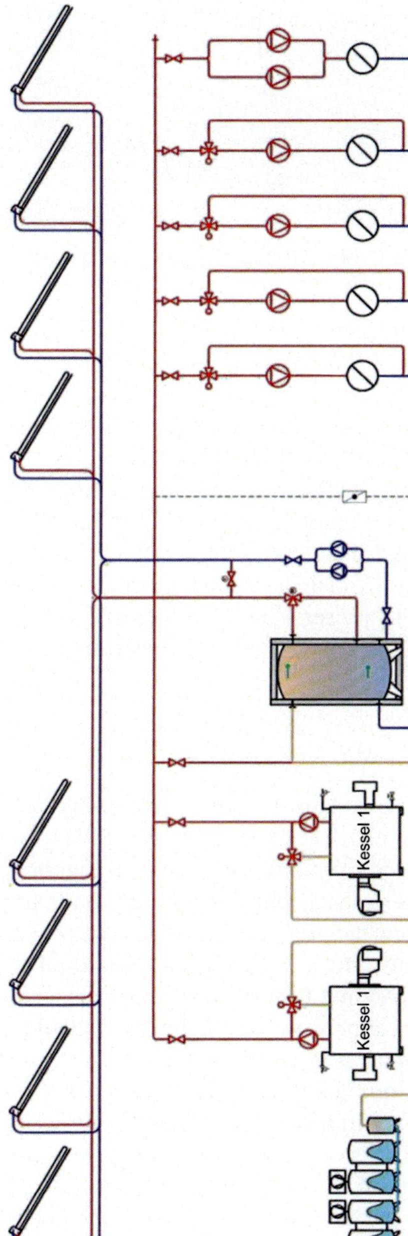

Figure 3.17 Integration of the evacuated-tube collector field into the heating network parallel to the existing boilers.
Source: Ritter XL Solar.

Figure 3.18 Air collector field for indirect drying of fruits with aperture area of 400 m^2. Source: Cona.

3.6.3 Drying

The company Alimentos Campestres S.A. in Guatemala City uses solar air collectors in combination with rock bed storage for continuous drying. The plant shown in Figure 3.18 can dehydrate up to 3000 kg of fresh fruits at a time. The collectors consist of wooden frames, white solar security glazing, and an absorber made of special aluminum profiles to ensure high heat conductivity. Ventilators suck the air through the collectors and into the horizontal drying chambers. Indirect drying (no direct sunlight) allows the product to preserve color, taste, and vitamins. High hygienic standards are met, because filters are applied to prevent the product from getting in contact with dust or insects.

3.6.4 Mining

During past years, mining industries have installed several very large-scale SPH plants, and more projects are being planned. The currently largest SPH system worldwide is the Minera Gabriela Mistral, pictured from bird's-eye view in Figure 3.19.

Mining needs very high investments, which makes the sector very risk sensitive and conservative about applying new technologies. However, there is a need for reduction of CO_2 emissions, because customers increasingly care about the carbon footprint caused by their supply chain. Adding to that, the sector sees an important advantage of SPH in the fact that solar heating costs are highly predictable over the lifetime of an SPH plant, which highly reduces investment risks.

Minera Gabriela Mistral is producing copper. There, energy is mainly needed for breaking the ore and transporting it, for pumps, as well as for electrolytic copper winning. Copper mines are usually very remote, so that energy supply has to be done by tank trucks delivering diesel or oil. By electrolysis, copper of high purity is deposited on plates. For this electrolytic purification, a copper solution is being heated in a bath to between 46 and 51 °C. Prior to the installation of the SPH plant, the mine used 8000 m^3 of oil per year to produce 120,000 tons of copper.

Figure 3.19 Flat-plate collector field of 39,300 m^2 at Codelco's "Minera Gabriela Mistral" in the Atacama Desert, Chile.
Source: Sunmark Solutions.

The SPH plant at Minera Gabriela Mistral has an aperture area of 39,300 m^2 and a heat storage of 4000 m^3. It was commissioned in October 2013 and has a thermal peak power of 27.5 MW_{th}. The system delivers heat for electrolytic copper winning to the Gaby copper mine, which belongs to the state-owned mining company Codelco.

The plant is located 100 km south of the city of Calama, in a region with almost no rainfall and regular sandstorms. Regardless of the harsh environment, the Atacama Desert is predestined for SPH, because the high altitude of between 2000 and 4000 m above sea level leads to 90% of very clear days. Additionally, the Atacama Desert is located near the equator between 18° and 25° southern latitude, so that there is only very little annual variation in irradiance. The primary (collector) loop of the plant works at 85 °C flow and 55 °C return flow, the secondary loop at 80/60 °C. The 80 °C flow temperature is needed to heat the baths. The secondary solar loop is directly connected to the boiler heating network. Due to the storage, the SPH system supplies heat for 24 h. The collectors are cleaned by an innovative automated dry cleaning device, and their tilt is seasonally adjusted to ensure maximum solar gains. The system annually delivers 50 GWh of heat, with an extremely high specific solar gain of 1272 kWh/m^2a. It covers 85% of the energy needed for electrowinning, equal to savings of 20,000 L of oil per day or 250 road tankers annually. The remaining heat demand is delivered by the existing oil burner.

Sunmark Solutions installed Minera Gabriela Mistral as a turnkey project. The project was realized by the project developer Energía Llaima SpA, who organized financing of the plant. Energía Llaima is contracted by Codelco, which only buys the solar heat actually produced. This way, Codelco did not have to take any risks, either technically or economically, and they did not have to put any effort into realization of the project (Slavin, 2015).

3.7 Future trends and research and development needs

The awareness of industrial and commercial companies for energy efficiency will further increase over the next few years, due to highly volatile energy prices, consumer request, and legal obligations in many countries, so process efficiency and waste heat recovery can be expected to play a bigger role in the future. Thus, SPH for applications below 60 °C will increasingly compete with waste heat recovery, which often can be realized at lower costs.

Generally, SPH will face increasing competition with photovoltaics (PV), if PV electricity costs continue to decrease faster than SPH generation costs. PV combined with heat pumps will be a challenge for low-temperature SPH wherever waste heat is available and area availability is not an issue. Competition with heat pumps can also be relevant for temperatures above 100 °C, because high-temperature heat pumps are technically improving.

Nevertheless, many promising future trends for SPH are observed. Market growth of low-temperature SPH is expected in agriculture and services (e.g., car wash), in which no waste heat is available. Technically modified thermal unit operations with reduced process supply temperatures may also create further opportunities for low-temperature SPH. Market deployment in small and medium enterprises is expected to continue, because their roof area is often restricted and long-term sustainable investments are possible even though the amortization times are above 3 years. In addition, the trend toward locally produced indigenous systems, as observed for solar drying or for the focusing systems in India, will go on. In these cases, SPH is not only the cheapest solution; it also offers heat supply security and creates local jobs.

Direct solar thermal steam generation with integration into existing steam networks is a very favorable application for SPH and is also expected to grow further in the future. These concepts can be highly standardized and SPH generation on supply level often tolerates future changes in the heat demand of single processes. The same considerations apply for direct integration of highly efficient stationary collectors into hot water networks.

At large-scale SPH applications, the high initial investment currently often results in a high share of capital costs and thus in long amortization periods. For financially demanding industries, contracting has proved to be a suitable measure to overcome these barriers. It has been successfully demonstrated that investor capital can make the realization of large-scale plants possible, if the technology is trusted (cp. e.g., Codelco mine). Thus, contracting is considered to be a key element for future growth of the SPH sector.

Based on considerations similar to the ones above, needs for increased research and development were identified in the framework of IEA SHC Task 49, the European Technology Platform on Renewable Heating & Cooling (RHC-Platform, 2014), and by others. Research and development should focus on:

- Medium-temperature collectors with improved efficiency, self-carrying and modular collector structures for large-scale collector areas on industrial roofs, as well as improved reflector materials for concentrating collectors

- Easy-to-handle planning and simulation tools for SPH
- Sector-specific, standardized solar heat integration concepts (branch concepts)
- Conception, realization, and dissemination of case studies for various industry sectors, process integrations, and locations (climate zones)
- Demonstration and dissemination of SPH cogeneration and polygeneration concepts, e.g., producing process heat, process cold, and electricity with high overall efficiency,
- Training and awareness raising
- Neutral third-party monitoring of highly promising demonstration projects to increase trust in SPH technology (e.g., of investors in SPH contracting)

Policy options to achieve higher market penetration are raising awareness for the benefits of SPH within industry clusters, providing financing mechanisms for upfront costs and subsidies for SPH instead of fossil fuels.

Sources of further information

EU-IEE Project SoPro. www.solar-process-heat.eu.

Frank, E., Mauthner, F., Fischer, S., 2015. Overheating Prevention and Stagnation Handling in Solar Process Heat Applications. Available from: http://task49.iea-shc.org/.

IEA SHC. International Energy Agency Solar Heating and Cooling Program. www.iea-shc.org.

IEA SHC Task 49. Solar Heat for Production and Advanced Applications. http://task49.iea-shc.org.

RHC-Platform. European Technology Platform on Renewable Heating and Cooling. www.rhc-platform.org.

References

CSH India, 2015. Concentrated Solar Thermal Heat. Available from: http://www.cshindia.in/ (08.04.15).

Häberle, A., 2012. Concentrating solar technologies for industrial process heat and cooling. In: Lovegrove, K., Stein, W. (Eds.), Concentrating Solar Power Technology. Woodhead, Oxford, pp. 602–619.

Helmke, A., Hess, S., 2015. Classification of solar process heat system concepts. In: Muster, B. (Ed.), Integration Guideline. Guideline for Solar Planners, Energy Consultants and Process Engineers. IEA SHC Task 49, pp. 61–76. Available from: http://task49.iea-shc.org/publications (25.05.15).

Hess, S., Oliva, A., 2010. Guide to Solar Thermal System Design for Selected Industrial Processes. EU IEE Project SoPro. Available from: www.solar-process-heat.eu/guide (01.03.15).

Hess, S., Oliva, A., Hermann, M., Stryi-Hipp, G., Hanby, V., 2011. Solar process heat – system design for selected low-temperature applications in the industry. In: Solar World Congress 2011. Kassel.

Hess, S., 2014. Low-concentrating, Stationary Solar Thermal Collectors for Process Heat Generation (dissertation). De Montfort University, Leicester, UK. Available from: http://www.reiner-lemoine-stiftung.de/en/pdf/dissertationen/Diss_Hess_Final_view_sh.pdf (03.05.15).

IEA, 2012. IEA Technology Roadmap Solar Heating and Cooling. International Energy Agency, Paris. Available from: http://www.iea.org/publications/freepublications/publication/Solar_Heating_Cooling_Roadmap_2012_WEB.pdf (01.03.15).

IRENA, 2015. Solar Heat for Industrial Processes. Technology Brief. International Renewable Energy Agency. Available from: www.irena.org/Publications (01.04.15).

Kalogirou, S.A., 2003. The potential of solar industrial process heat applications. Applied Energy 76 (4), 337–361.

Lauterbach, C., Schmitt, B., Jordan, U., Vajen, K., 2012. The potential of solar heat for industrial processes in Germany. Renewable and Sustainable Energy Reviews 16 (7), 5121–5130.

Mauthner, F., Hubmann, M., Brunner, C., Fink, C., 2014. Manufacture of malt and beer with low temperature solar process heat. Energy Procedia 48, 118–1193.

Muster, B. (Ed.), 2015. Integration Guideline. Guideline for Solar Planners, Energy Consultants and Process Engineers. IEA SHC Task 49. Available from: http://task49.iea-shc.org/publications (08.04.15).

Norton, B., 1992. Solar Energy Thermal Technology. Springer-Verlag, Heidelberg, Germany.

POSHIP, 2001. The Potential of Solar Heat for Industrial Processes. Available from: http://www.solarpaces.org/images/pdfs/poship_final_report.pdf (01.03.15).

SHIP-plants, 2015. Database for applications of solar heat integration in industrial processes. Available from: http://www.ship-plants.info/ (28.02.15).

Solar Keymark. Collector Test Certificates. Available from: http://www.solarkeymark.dk/CollectorCertificates (01.03.15).

SRCC. Solar Rating and Certification Corporation. Collector Test Certificates. Available from: http://www.solar-rating.org/ratings/index.html (05.03.15).

Sunfocus, January–March 2015. A Quarterly Magazine on Concentrated Solar Heat, 2 (3). Available from: http://mnre.gov.in/file-manager/UserFiles/Sun-Focus_January-March-2015.pdf (01.03.15).

Slavin, A., 2015. Renewables are customizing PPAs for mining industry, but more dialogue necessary. Renewable Energy World.com. Available from: http://tinyurl.com/pw7k3n3 (08.04.15).

Tamme, R., Bauer, T., Buschle, J., Laing, D., Müller-Steinhagen, H., Steinmann, W.-D., 2008. Latent heat storage above 120 °C for applications in the industrial process heat sector and solar power generation. International Journal of Energy Research 32, 264–271.

UNIDO (United Nations Industrial Development Organization), 2011. Renewable Energy in Industrial Applications. An Assessment of the 2050 Potential. Available from: http://www.unido.org/fileadmin/user_media/Services/Energy_and_Climate_Change/Energy_Efficiency/Renewables_%20Industrial_%20Applications.pdf (01.03.15).

Vannoni, C., Battisti, R., Drigo, S., 2008. Potential for Solar Heat in Industrial Processes. IEA SHC Task 33/IV. Available from: http://task33.iea-shc.org/ (01.03.15).

Weiss, W., Mauthner, F., 2014. Solar Heat Worldwide: Markets and Contribution to the Energy Supply 2012. Gleisdorf. Available from: http://www.iea-shc.org/solar-heat-worldwide (01.03.15).

RHC-Platform, 2014. Solar Heating and Cooling Technology Roadmap. European Technology Platform on Renewable Heating and Cooling. Available from: www.rhc-platform.org (01.03.15).

RHC-Platform, 2011. Common Vision for the Renewable Heating & Cooling Sector in Europe. European Technology Platform on Renewable Heating and Cooling. Available from: www.rhc-platform.org (05.03.15).

Deep geothermal energy for heating and cooling

4

Ruggero Bertani
Enel Green Power, Italy

4.1 Introduction

4.1.1 Geothermal resource classification

Geothermal energy is a clean and renewable energy source deriving from the exploitation of the heat flux reaching the surface from the deeper part of the Earth's crust. The amount of such thermal energy is very large, and it originates from the natural heat of the Earth primarily due to the decay of the radioactive isotopes of uranium, thorium, and potassium. At the base of the continental crust, temperatures range from 200 to 1000 °C, and the heat is transferred toward the surface, mostly by conduction, causing an average rise in temperature of 25 to 30 °C/km. This is called the geothermal gradient. In most promising locations, typically along fault lines or tectonically active zones, the value can be up to 5 to 10 times the average, with very significant increases of temperature even at shallow depths.

However, due to several technical, logistical, and thermodynamical limitations, only a fraction of the heat can be utilized. In a more rigorous sense, geothermal energy is only that part of geothermal heat that can be exploited through drilling. A detailed estimation of the heat stored 3 km deep under the continents by the Electric Power Research Institute (EPRI) (1978), considering separately the average geothermal gradient for normal geological conditions and the diffuse geothermal high-enthalpy regions, evaluated a total amount of available heat about 42×10^6 EJ, and the average heat flow on the surface about 65 mW/m^2 globally (Pollack et al., 1993), as shown in Figure 4.1.

Traditionally, geothermal energy was considered exploitable only in areas where naturally occurring water (or steam) is found concentrated at depths less than 3 km, and at temperatures above 30 to 180 °C (Cataldi, 1999; Fridleifsson et al., 2008). This view is changing with the penetration in the market of technologies that can economically utilize lower-temperature resources (between 100 and 180 °C for electrical production) and the emergence of ground-source heat pumps using the Earth as either a heat source or sink, depending on the season (Curtis et al., 2005). However, the basic principle is very simple: water from the geothermal system and the cooled water from the surface are transported using production and injection wells, and recovered as steam or hot water or both. The utilization of geothermal energy could be electrical production, direct application, and both in cascade, depending on fluid temperature.

Renewable Heating and Cooling. http://dx.doi.org/10.1016/B978-1-78242-213-6.00004-7

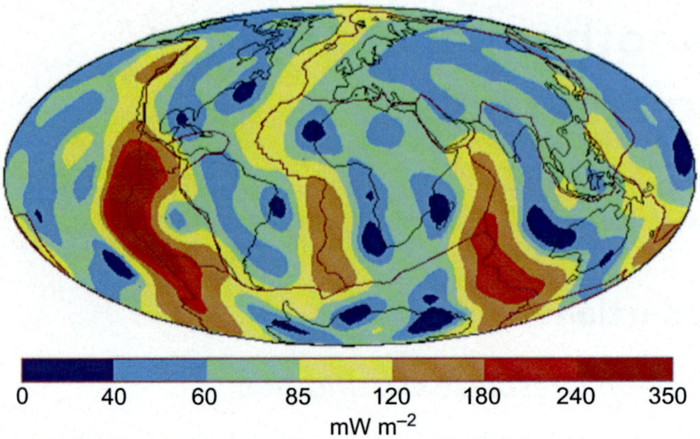

Figure 4.1 World geothermal heat flux (Intergovernmental Panel on Climate Change (IPCC), 2011).

4.1.2 Utilization of deep fluids

A detailed discussion of the utilization of the shallow geothermal system through heat pump will be presented in Chapter 5 of this volume. We will focus our discussion only on deep-fluid exploitation.

A typical geothermal reservoir contains brine (a hot, concentrated, and saline water solution, enriched in substances leached from crustal rocks due to its circulation in an area of anomalously high heat flow, chlorides of Na, K, and Ca, dissolved metals or steam, with presence of fractions of dissolved gases (CO_2, H_2S, hydrocarbons). Its energy content is described in terms of enthalpy or temperature. High (>180 °C) and medium (>100 °C) temperature resources from hydrothermal reservoirs are used for electrical production with different technologies, depending also on the geothermal fluid and its chemical and gas content. Low- to medium-temperature resources can be used directly as heat in district heating, for agriculture (greenhouses) or aquaculture, industrial process heating, and in spas. The geothermal exploitation of a deep system is dominated by two technological aspects: the extraction of heat energy using the fluid as carrier, and its transformation into an usable form.

The first item is realized through a geothermal well, in the depth range of 1 to 3 km, in which naturally occurring geothermal fluids are extracted from the Earth, either artesian or pumped, and later either reinjected or (less frequently) rejected. A detailed discussion of the different underground components will be given in the next chapter.

The transformation is strongly dominated by the thermodynamic evaluation of the possible work and the conversion efficiency. For electrical generation, the technologies are mostly linked to the temperature and pressure of the geothermal fluid (Figure 4.2). The most efficient one is the "direct steam turbine," using high-temperature steam resources directly to generate electricity, with very good efficiency (20–25%) and relatively modest power plant cost. An example of this technology is the Larderello geothermal field in Tuscany (Italy). The most common high-temperature system, with a mix of brine and steam, is the "flash steam plant"; the steam is first separated from the liquid and then expanded in a turbine. It is possible to divert the hot separated

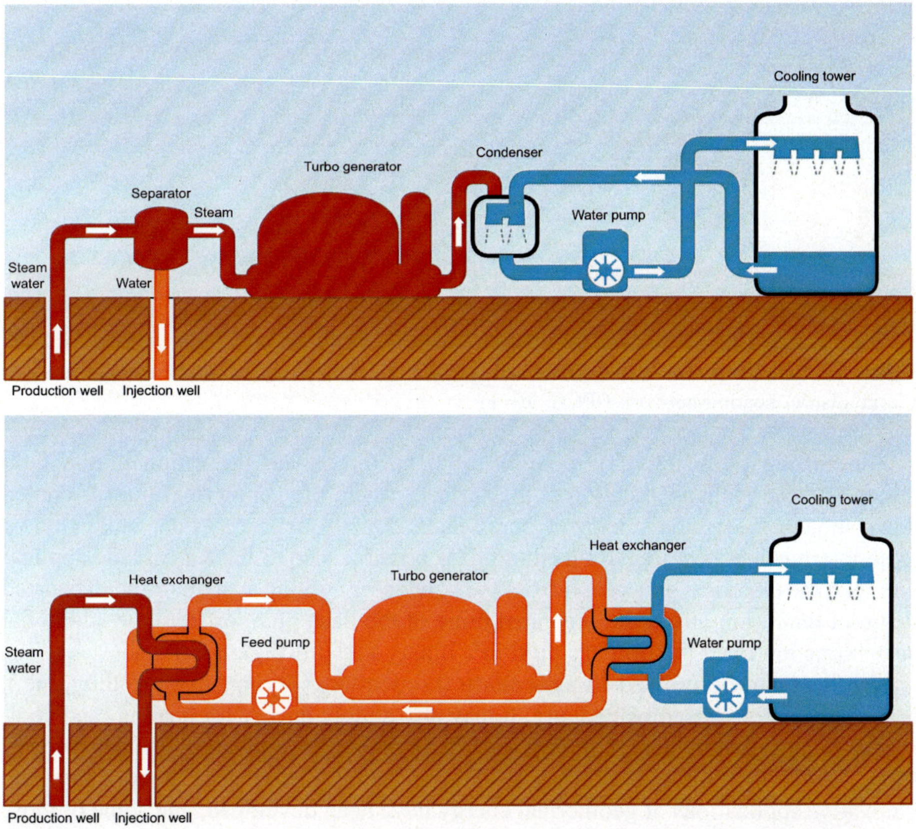

Figure 4.2 Schematic view of a steam condensing plant (top) and a binary plant (bottom). The difference between the dry steam and the flash plant is the presence of the separator at the well head for flashing the geothermal fluid and collecting a steam flow to the turbine (IPCC, 2011).

brine to heat applications (cascade generation) or secondary low-temperature electrical generation through a binary plant. Sometimes, it is possible to have a "double-flash system," in which the hot brine passes through successive separators each at a subsequently lower pressure. The two steam flows are directed to different turbines to maximize the overall efficiency. In all of the aforesaid options, the steam is condensed after the expansion in the turbine and the remaining fraction after the evaporation in the cooling tower is reinjected. The advantages of a better utilization of the geothermal resource can compensate the overall cost increase. Finally, the "binary cycle technology" is the most suitable for the temperature range 100–180 °C, with an efficiency strongly dependent on the resource temperature in the range 7–12%, higher than a flash plant in similar conditions; it separates into two loops the geothermal brine from a secondary fluid that is vaporized, then expanded through the turbine, condensed through an air- or water-cooled condenser, and pumped back to the heat exchanger to be vaporized again. The geothermal fluid is totally reinjected, without any evaporation losses (Dickson and Fanelli, 2003).

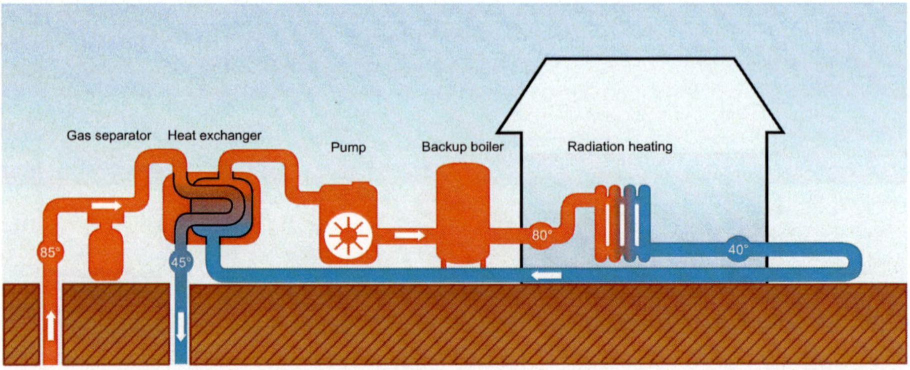

Figure 4.3 Schematic view of a typical heating application; the end user can be a single house or a large district heating network (IPCC, 2011).

For heating purposes, it is possible to have direct-heat utilization of the fluid through a heat exchanger station at the surface, where the geothermal fluid is cooled down though a secondary loop of pure water, directly usable as final utilization or as an intermediate heat transfer medium. It is possible also to have a cascade application with the electrical generation from the condenser cooling system of the exhausted steam of the turbine from the separated brine of the flash plant or from the additional heat extraction from the primary fluid in a binary system (Figure 4.3).

The heat can be utilized in agriculture (horticulture, drying, fish-breeding, etc.), industrial processes, and balneology. It may also supply energy for heating (and cooling) districts and individual buildings, including both small and large schemes (offices, shops, residential houses, schools, greenhouses, bathing, etc.). A number of new and innovative applications of geothermal energy have been developed, and some of those have already been demonstrated (ice/snow melting, seawater desalination, etc.) that logically can be supplied by geothermal district heating systems.

The world applications of geothermal energy in the different technologies are listed in Tables 4.1 and 4.2 (Lund et al., 2010; Lund, 2008).

Table 4.1 **Electricity**

Category	Installed capacity GW	Number of plants	Average capacity MW	Percentage on the total installed capacity
Binary	1.5	264	5.7	13%
Single flash	4.8	155	31	41%
Double/triple flash	2.4	67	36	21%
Dry steam	2.9	63	45	25%

Table 4.2 Direct uses

Category	Utilized energy TJ_{th}/year	Percentage on the total utilization
Geothermal heat pumps	214,782	49%
Space heating	62,984	14%
Greenhouse heating	23,264	5%
Aquaculture pond heating	11,521	3%
Agricultural drying	1662	0.4%
Industrial uses	11,746	3%
Bathing and swimming	109,032	25%
Cooling/snow melting	2126	0.5%
Others	956	0.2%

4.2 Direct heat utilization technologies

4.2.1 Subsurface installations (well, submersible pumps)

The drilling processes and equipment utilized in deep geothermal wells are substantially similar to those developed for petroleum and water-well rotary drilling. However, the particular geothermal environment requires some significantly different practices to be adopted. A geothermal well is usually made in different sections, as indicated in Figure 4.4.

Because the thermal efficiency of utilization of geothermal fluid is not particularly high, especially in converting to electricity, a large flow rate is required, implying large-diameter production casings and liners (standard API 9 5/8″ diameter as production casing). A quite relevant difference with oil and gas wells is that all of the casings are usually run back and fully cemented to the surface, due to the high thermal stresses that require uniform cementation for avoiding stress concentration points. The typical drilling fluid is water-based bentonite mud treated, controlling the viscosity for the increase of temperature and pressure with additions of simple dispersants. Loss of circulation zones are usually plugged and controlled to reach the productive depth and running the production casing shoe.

In some situations, the reservoir pressure is high enough for triggering a self-producing well, and the fluid reaches the surface with a time-dependent flow rate, usually slowly declining, depending on several reservoir and well parameters (permeability, skin factor, inner casing diameter, well depth, scaling, reservoir pressure, interference effect with other wells, etc.). When the well productivity is no longer commercial, it is common to drill a new one or to execute maintenance activity (work over) to try to keep the production close to the design level.

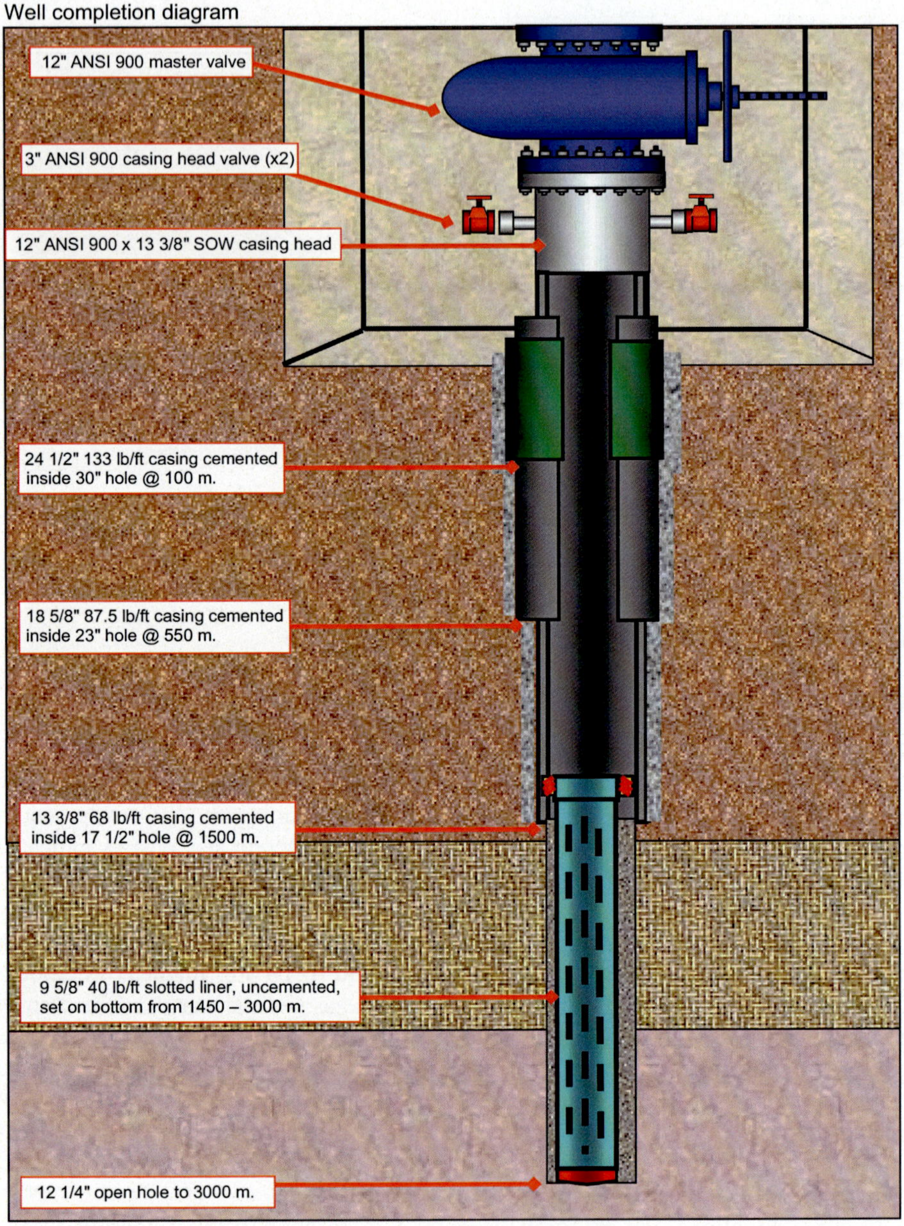

Figure 4.4 Typical well design.
Courtesy of Enel.

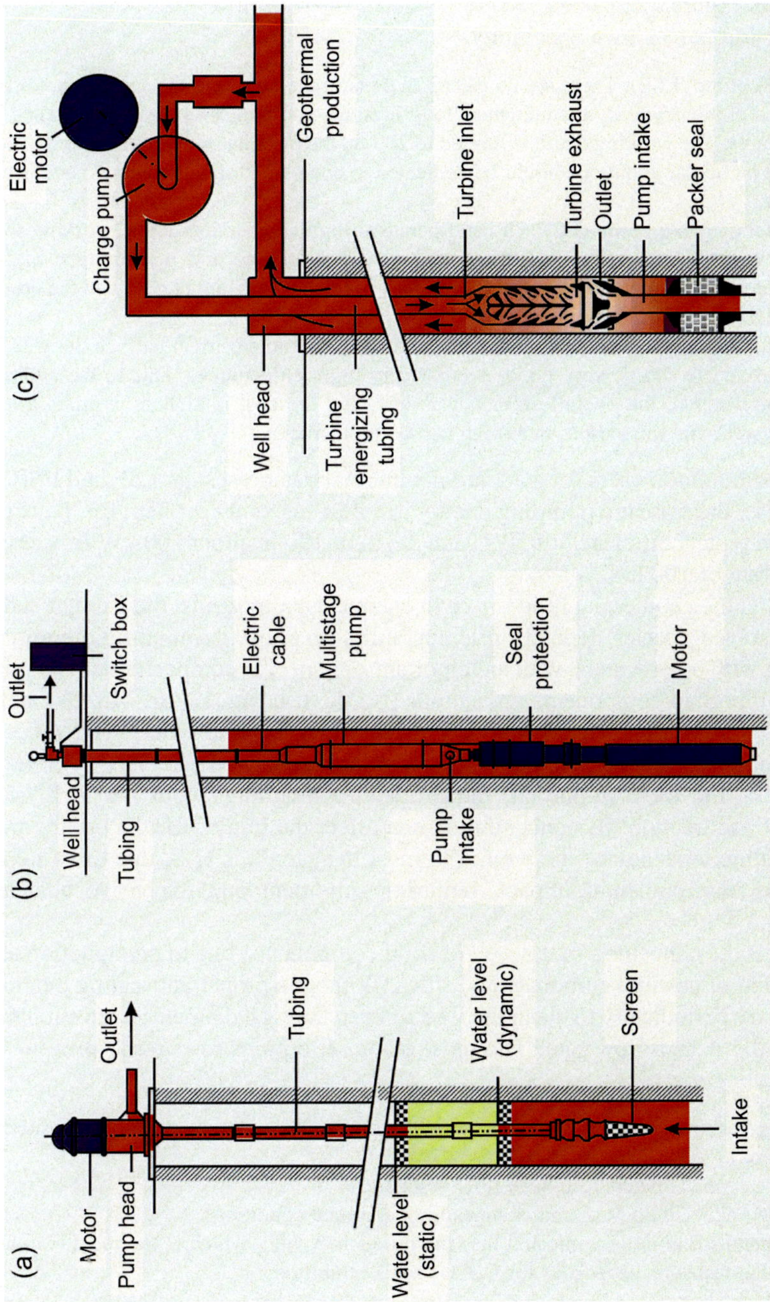

Figure 4.5 Solution for production well pumping (Ungemach, 2004). (a) Line shaft pump (LSP). (b) Electro-Submersible pump (ESP). (c) Hydraulic turbine pump (HTP).

When the flow rate is minimal or the water level in the well is below the surface, it is necessary to pump up the geothermal fluid. There are several technical solutions, as illustrated in Figure 4.5.

Each option has its own specificity:

1. Line shaft pump (LSP): There are no electrical parts in the hole, and the surface motor has a very good efficiency, easy maintenance, long lifetime, and attractive costs. It tolerates high temperatures. The working depth is limited to 200 m, the installation/removal activity is delicate, and particular attention should be dedicated to enclosing tubing, coatings, and bearing materials.
2. Electro-submersible pump (ESP): It can operate at high submersion depth, working with a high flow rate even in small-diameter casing. Good lifetime and performance in high-temperature environments. The pump has a lower efficiency and higher cost, and the problem of electrical insulation can be relevant.
3. Hydraulic turbine pump (HTP): As for the LSP, there are no electrical parts in the well, and the operative life can be very good, even at very high temperatures. Due to the additional energy conversion, the overall efficiency is low, and the cost is high; it requires a large-diameter well, and the packer anchoring can be problematic.

The most common users for standard geothermal businesses are LSP and ESP. For both of these, the parasitic pumping energy can be a relevant cost item for the exploitation. With typical efficiency of 50%, at a depth of 100 m about 200 kW is necessary for a flow rate of 100 L/s.

Generally, the geothermal fluid is rich in chemical components; the amount and nature of dissolved species depend on temperature, pressure, minimal-fluid equilibria, and mixing with other waters, with an important site-specific compositional variability. The most important components are cations (Na, K, Ca, Mg, Li, Sr, Mn, Fe), anions (Cl^-, HCO_3^-, SO^-, F^-, Br^-), non-ionic (SiO_2, B, NH_3), and others (Hg, As, some metals). The dissolved non-condensable gases include CO_2, H_2S, NH_3, N_2, H_2, and CH_4. CO_2 is the most important, quite often representing about the 90% of the geothermal gas fraction. Its concentration can affect the liquid pH, the boiling point, and the scaling tendency of the well. The most dangerous is H_2S, due to its toxicity and serious environmental impact, requiring important and expensive abatement measures.

Scaling is the deposition in the well of solid components, due to precipitation at the flashing point of calcium carbonate and silica. Without a proper antiscaling treatment, it is necessary periodically to clean the well to keep the well diameter large enough for sustaining the design flow rate. The most common options to control, prevent, and mitigate scaling are:

- Avoiding flashing in the well, keeping enough pressure through the operation of a submersible pump
- Enlarging pipe diameters to reduce flow velocities
- Avoiding undue shutdowns and changes in operational conditions
- Using chemical inhibitors, injected into the production well, which is the most commonly and routine measure, especially for $CaCO_3$ scale inhibition

Corrosion is the reduction of the integrity of metals exposed to hostile fluid thermochemical environments. It can be uniform or spot (pitting corrosion). The higher the temperature, the higher the corrosion rate, and of course it is strongly affected by the decreasing pH (i.e., increased acidity) and increasing oxygen concentration, which is a major corrosion agent, as well as the chloride component. Corrosion rates also increase with fluid velocity. The best option for corrosion mitigation is a suitable material selection and the utilization of chemical inhibition, requiring the realization of a suitable injection scheme and an adequate selection of inhibitor agents, as well as accurate monitoring and evaluation protocols. An example of a chemical injection setup is shown in Figure 4.6.

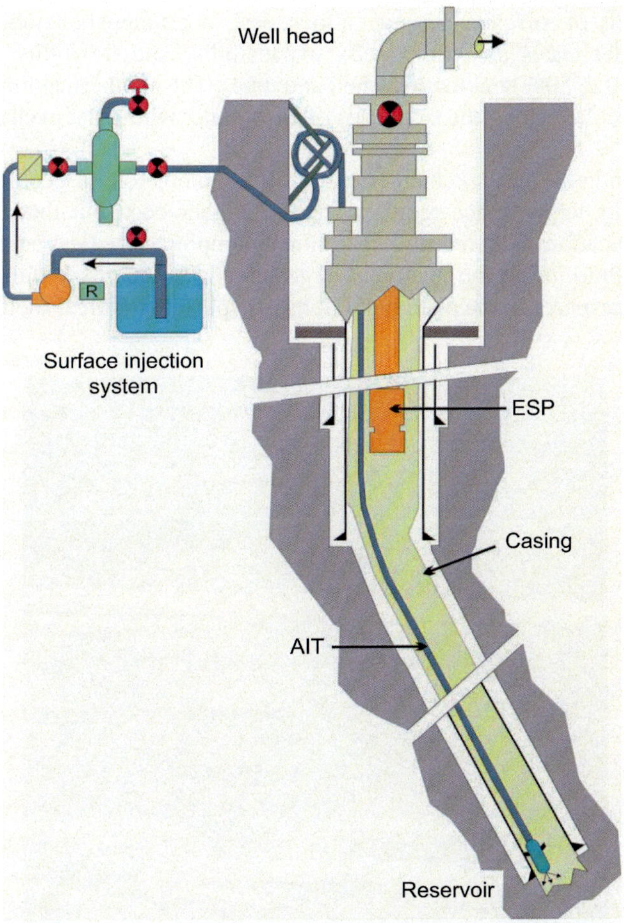

Figure 4.6 Solution for chemical injection (Ungemach, 2004).

4.2.2 Utilization

The utilization of the geothermal fluid, once it is delivered to the surface, is mainly made through a heat exchanger, which allows keeping separate the geothermal line from the secondary loop line, usually containing clean water, which is piped directly to the end user or to another heat-exchanging station, in case of long-distance transfer from the production zone to that of utilization. As already highlighted, heat utilization can be very different, and it is possible to cascade modes, using the thermal energy at different levels of temperature, according to specific needs (Rafferty and Gene Culver, 1998).

The heat transfer process is most often handled by a plate heat exchanger; there are two main configurations, bolted or brazed, but both of them exhibit excellent thermal performance (up to three/four times better than shell and tube systems), a relatively low cost in comparison with the other components of the geothermal cycle, the availability of a wide variety of corrosion-resistant alloys, ease of maintenance (tube cleaning and replacement clearances are eliminated), expansibility and flexibility in a compact design (from 10 to 50% smaller than shell and tube). The geothermal fluid and the secondary one circulate along the two faces of each plate, with a flat exchanging surface (Figure 4.7).

The shell and tube heat exchanger plays only a minor role in geothermal applications, due to its larger space requirement and higher cost, but that is suitable for very large applications with reduced operational pumping costs (less energy is required for moving a fluid inside the relatively large tube, in comparison with the small flat rectangular interspaces in the plate type). It has a typical cylindrical shell with multiple

Figure 4.7 Photo of a large heat plate exchanger.

Figure 4.8 Photo of a shell and tube heat exchanger.
Courtesy of Enel.

tubes inside. One fluid passes inside the tubes; the other fluid circulates on the outside of the tubes within the cylindrical shell. Heat is transferred from one fluid to the other through the walls of the tubes (Figure 4.8).

It should be highlighted, however, that for the binary plant application, in which a large flow rate is necessary, shell and tube are the most used solutions, as well as for the large district heating heat exchange primary stations.

There are several applications of geothermal heat after its transfer to the secondary loop cycle and to the end user: bathing, space and district heating, agricultural applications, aquaculture, and some industrial uses are the best known forms of utilization. Heat pumps are the most widespread, and they will be covered in detail in Chapter 5 of this paper.

- Space and district heating: these applicatiosn are highly capital intensive (but with relatively low operating expenses), with investment costs for:
  - production and injection wells
  - downhole and circulation pumps
  - pipelines
  - distribution networks
  - monitoring and control equipment
  - peaking stations
  - storage tanks

- Agricultural applications can be open field (using thermal water to irrigate or warm the soil) or most commonly as greenhouse heating, developed on a large scale in many countries. Using geothermal heat can considerably reduce the overall operating costs. The heating can be accomplished by:
  - forced circulation of air in heat exchangers
  - hot-water circulating pipes or ducts located in or on the floor
  - finned units located along the walls and under benches
  - a combination of these methods on air movement
- Aquaculture (the controlled breeding of aquatic forms of life in warm water) is growing in importance, due to an increasing demand for fish. By maintaining an optimum temperature it is possible to double the reproductive cycle in some very common species.
- The entire temperature range of geothermal fluids, whether steam or water, can be exploited in several industrial applications or food transformation application (milk treatment):
  - process heating
  - evaporation
  - drying
  - distillation
  - sterilization
  - washing
  - deicing
  - salt, mineral, gas extraction, etc.

4.3 Resource assessment and economics

4.3.1 Exploration technologies

The drilling cost is one of the most important cost items for geothermal activity. It is very important to minimize the risk of not achieving a good result from the activity. The most important measure is to invest in surface exploration activities, to get a valuable set of information, which can be extremely important for better well location and targeting.

It should be out of the scope of the present paper to enter in many details of all the available surface exploration techniques, but nevertheless we will provide a very general overview.

The exploration has as its primary target the identification of a geothermal field, its size, type, and major characteristics. It is important to locate productive zones, to evaluate the heat content of the fluid and the environmentally sensitive parameters.

- Geological and hydrogeological studies: These are extremely important, and it is almost impossible to plan a geothermal development without a solid geological evaluation of the area. They provide the location, the broad range extension of the interesting zone, the location of the wells, the deep fluid circulation, and the presence of natural recharge. The duration and cost of the entire project can be appreciably reduced by a good preliminary geological survey.
- Geochemical surveys (including isotope geochemistry): They are conducted on the surface water, spring, and natural manifestations. The cost is relatively low compared to other more sophisticated methods, and it should be conducted as much as possible with the target of

estimating the underground water temperature, the homogeneity of the water supply, the chemical characteristics of the deep fluid, and the source of recharge water.
- Geophysical surveys: Through these technologies it is possible to obtain from the surface information on several physical parameters:
  - temperature (thermal survey)
  - electrical conductivity (electrical and electromagnetic methods)
  - propagation velocity of elastic waves (seismic survey)
  - density (gravity survey)
  - magnetic susceptibility (magnetic survey)

  The most important methods nowadays are seismics, gravity, and magnetotelluric, which can produce a 2D or 3D reconstruction of shape, size, and depth of the geothermal system.
- Thermal techniques: Through the determination of geothermal gradient and terrestrial heat flow, it is possible to provide an evaluation of the temperature at the top of the reservoir, using the extrapolation of the measured gradient at the expected depth as inferred by the geological or geophysical surveys.

An optimal exploration technique does not exist. All of them are (more or less) expensive, and they exhibit relative advantages and disadvantages; a proper selection should be done case by case, having in mind also the expected total budget of the project. The first step should be a critical review of all the existing geological, geophysical, and geochemical data, with a reconnaissance phase, in which a region of interest could be focused upon, then a prefeasibility and feasibility study will follow, in which a greater level of detail (and expenses) will be reached, concentrating on the most promising zones. Periodic reassessments of the program would be necessary, to ideally eliminate any operations no longer necessary and insert others, with a delicate balance between information and expenses, with the objective of reduction of mining risk without overburdening the total project costs.

4.3.2 Geothermal costs

The cost structure of a geothermal project is dominated by three major items: exploration, drilling, and facility construction (Hance, 2005; Kagel, 2006).

As already discussed in Chapter 3.1, in the exploration phase the geothermal resource is identified and located, at distinct levels of depths, and the overall energy content is estimated. It comprises different surface exploration surveys and drilling of slim holes and/or geothermal gradient wells for thermal evaluation. In a typical project the following costs could be estimated (not all the surveys indicated in the list should be planned):

- *Assessment of existing data*, acquisition of satellite images, etc.: Minimal cost, up to a few thousand euros, according to the amount of existing data, the time availability, the size of the area, the availability of an electronic data set the cost for buying from specialized services the same specific data set, etc.
- *Geological survey*, hydrogeologic studies, compilation of geological map and in situ evaluation of geothermal features: It is also a low-cost activity, which requires a few man-months of desk work and at least one field trip for direct data collection and observation. Cost is dependent on the logistical expenses for the location and size of the geothermal site and

the needed amount of geologist manpower and equipment (GIS station and dedicated software); it can be considered usually below 100,000 Euro in total.

- *Geochemical survey* and laboratory analysis of the collected samples; also for this task the same consideration for the geological part can be applied: Low-cost item, with one field trip for sample collection, a laboratory phase, and an interpretation one. A standard set of laboratory measurement costs in the range 200–600 Euro per sample, with an overall cost of the entire activity below 100,000 Euro.
- *Geophysical surveys–gravity*: Often used in conjunction with other geophysical methods. A standard number of one station per km^2 is enough for the majority of situations, extending outside the area to 2 stations per km^2, necessary for achieving good resolution inside the interesting zone, avoiding boundary effects; the typical cost for data acquisition and elaboration of about 100–200 Euro each station; an expected total of 100,000–200,000 Euro for a detailed gravity acquisition, elaboration, and interpretation is reasonable.
- *Geophysical surveys–magnetotelluric*: This exploration technique is quite effective for a rough evaluation of the expected reservoir boundaries (in size and thickness), and interpretation should be done with all the other available information, using also the results of the first well for a proper tuning of the expected correlation of inferred resistivity at depth and geological layers. A typical data acquisition density is one station per km^2 at a cost of 1000–1500 Euro each; an expected total of 100,000–200,000 Euro for a detailed magnetotelluric acquisition, elaboration, and interpretation is standard.
- *Geophysical surveys–seismic*: It is the most effective tool for realizing a detailed 2D or 3D model of the subsurface, achieving a good resolution and acquiring valuable information on the possible well target in depth and location. Unfortunately, seismic is expensive, due to the cost of realizing the seismic data acquisition network and activation with vibroseis or explosives over a large area. One km of seismic acquisition, elaboration, and interpretation can cost 10,000–20,000 Euro, depending on logistic and topography, and for a good coverage of a standard area the cost could be about 200,000–400,000 Euro. With a 3D approach, the cost is about 50,000 Euro per km^2, reaching easily 300,000–500,000 Euro of expenses.
- *Geophysical surveys–others*: All the other techniques for data acquisition and elaboration (resistivity, natural seismicity, magnetic, etc.) are in general low cost and do not significantly affect the total exploration budget.
- *Drilling slim holes*: Slim-hole drilling is considered part of the surface exploration activity, achieving temperature and geological information of the shallow part of the system, with a maximum depth of 1 km. Holes are usually drilled with the continuous coring technique, and a typical well cost could be between 300,000 and 900,000 Euro, strongly dependent on the logistics, accessibility, availability in the country of drilling companies and materials, topography, permitting and local regulations, land cost for acquisition of drilling pad, etc.
- *Drilling gradient wells*: A very shallow gradient hole can be drilled for a thermal measurement of heat flux and temperature in the coverage terrain (a few hundreds of meters), which can be used for inferring information about the deep systems. Their cost is relatively cheap, about 100,000–300,000 Euro.

Site development covers the drilling activities necessary to the commercialization of a geothermal resource. Recently, drilling costs have increased significantly because of competition from oil and gas activities for drilling-rig availability. They are dominated by several factors, such as the permeability of the rock, depth of the reservoir, chemistry of geothermal fluid, resource temperature and pressure, logistics and accessibility, as well as the general permitting and regulation issues.

Table 4.3 **Range of drilling costs**

Depth (m)	Cost minimum (k Euro)	Cost maximum (k Euro)
1000	1500	2500
2000	3500	4500
3000	5000	6500
4000	5500	7000

A standard geothermal production/reinjection well cost is slightly nonlinear with depth (in meters), following the approximate formula (overestimating a very shallow well, less than 1–1.5 km):

$$\text{Cost (Euro)} = 800,000 + 2000^{*}\text{Depth} - 0.127^{*}\text{Depth}^{2}$$

The fixed cost of 800,000 Euro is due to the drilling pad and rig moving. This value is only approximated, depending on the rig availability, its size, logistics, and number of wells drilled from the same pad.

The linear component of the cost is due to the material, personnel, rig per day or per meter, services: all these items are proportional to time or depth, which are roughly linearly correlated. There is a small correction for greater depth, in the square power term. It is due to the major cost in the first part of the well due to the number of casings and cementing operations in place, in comparison with the deeper part of the well, usually in open hole, or slotted liner not cemented. The following range of values (Table 4.3) could be used as a preliminary indication of the expected investment for the drilling part of the project.

The total cost will depend on the total number of wells (necessary for achieving the required flow rate and thermal output) and their success ratio, usually 70–80% in the developing phase.

In the case of nonartesian flow, or if a higher flow rate is required, the utilization of a downhole pump is necessary. Generally, this equipment has a rather short operative life, 3–4 years approximately, and a relevant parasitic consumption, with an important impact on operations and maintenance (O&M), due to the replacement and operational costs. A typical investment cost for a standard downhole pump is about 500,000 Euro.

Finally, the last cost item of the site development phase is the surface installation cost. It is almost impossible to evaluate the expected investment, due to the relative impact on the end-user application side: length of pipeline, cost of utilization system, and administrative and permitting expenses (which can be relevant, in case of a large district heating network).

It is possible only to evaluate a draft heat exchanger cost per thermal energy unit supplied, closing the geothermal loop at that boundary, which is also the delivery point of the heat supplier to the end user. An average cost is about 10,000 Euro for each Gcal/h unit (1.16 MW_{th}).

4.3.3 Risk management

The entire geothermal process from the initial exploration to its exploitation phase is considered a high-risk investment. This risk is associated with the unknown nature of the underground resource and at the standard difficulties that can arise from a complex industrial application. An important factor on the risk aspect is the time lapse from the initial phase till the commissioning, which can be long, up to 5 to 10 years. This lag can discourage modest-profile investors, also due to the higher risk of the initial phases of the project (preliminary surveying and exploration phases), but the expenditures are relatively low at that stage. The test drilling phase requires an increased level of expenditure, still at a high level of uncertainty. The majority of unsuccessful projects stop at this step. Risk mitigation funds have been established in some countries to help overcome this difficult phase. However, the best insurance is a good detailed and exhaustive exploration phase. The utilization of several associated techniques minimizes the uncertainty in ascertaining resource temperature, depth, productivity, sustainability, and well locations. A good interpretation of exploration data enables development of an initial conceptual model of the geothermal system, which is of fundamental importance for success of the entire project.

Significant risks are associated with drilling activities, as a function of the drilling conditions (logistical, timely availability of equipment and services, rock type, borehole competency and pressure conditions, chemical and gas presence, and so on); the risks are of two major types: it could be possible to realize a good well, but without finding an exploitable fluid for several different considerations, or the impossibility of concluding the drilling phase due to technical problems (casing collapse, blow out, permanent loss of materials in the hole, etc.). The first can kill the project immediately, in case the bad results would be extrapolated to the entire inferred geothermal system, as "lack of exploitable resource." The second can be only a "bad luck event," which can result only in a loss of the capital invested for the well.

It should be important also that, after a successful drilling phase, the geothermal resource would be able to feed consistently and reliably to the plant during its expected operative life (20 years, typically). Resource degradation can anticipate the decline in the production rate or exhibit a premature cooling, as well as an adverse chemical effect. It is common to plan additional drilling of "makeup wells" during the life of the project for keeping the production level stable and not penalizing the evaluated internal rate of return (IRR) of the project.

4.4 Case history of major district heating

4.4.1 Paris (France), a very large and long-lasting application

One of the largest district heating systems is in Paris, in the so-called Dogger aquifer, hosted in limestone rock between 1500 and 1800 m depth at 70 °C, with more than 60 doublets and triplets of wells for heating purposes. The development had its historical maximum in the 1970s and 1980s due to the oil crisis, and nowadays it is in an expansion phase, with several new projects at different stages of planning and realization.

Today, the geothermal district heating network of the Paris basin has the largest concentration of wells in the same aquifer in the world: for 40 years of geothermal exploitation in the Paris basin, more than 120 wells have been drilled to exploit the Dogger reservoir.

The operation of the system is based on the following three modules:

- The geothermal supply module, made by a doublet/triplet of wells for production and reinjection, the pumps, surface piping loop, and electrical/hydraulic control devices
- The demand module, from the end-user side, consisting of heaters connected to the distribution grid
- The backup module, made by fossil fuel boilers, interconnected with the geothermal heat exchanger station

The savings from the geothermal heating can be evaluated as 120,000 ton CO_2/year, with 250 MW_{th} installed and a heat delivery of 1500 GWh_{th}/year for 170,000 buildings. More than 70% of the entire project cost is for the distribution grid (Figure 4.9).

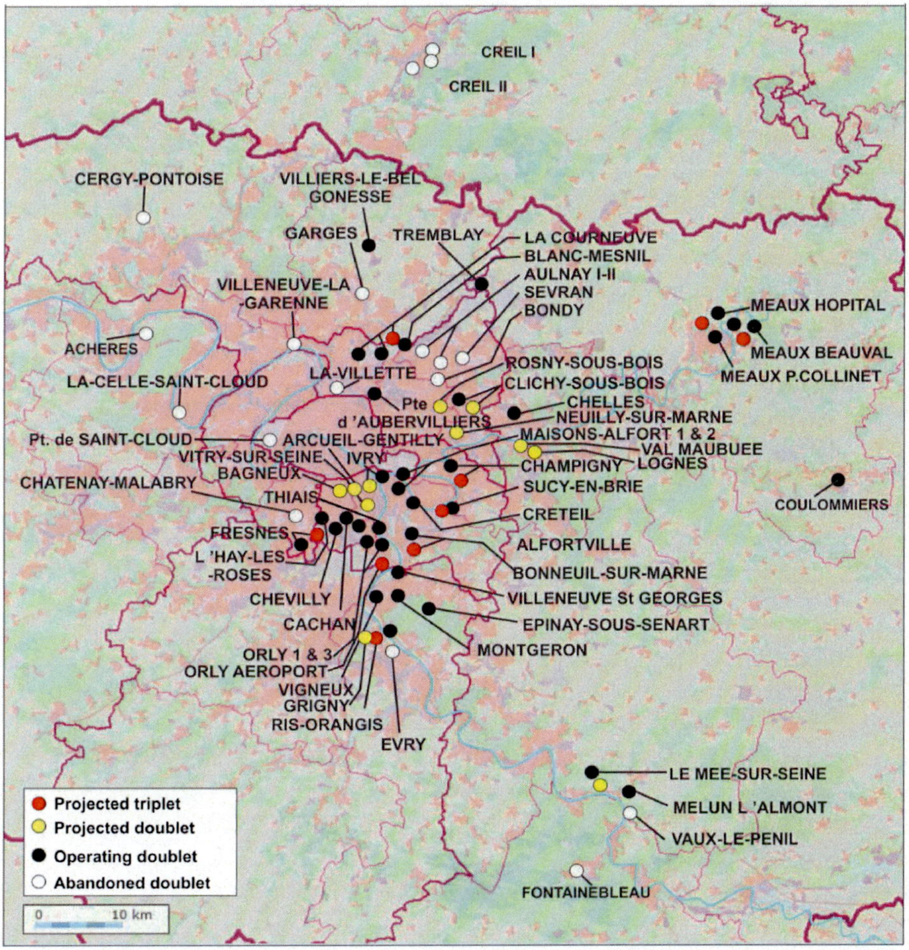

Figure 4.9 Paris district heating (Ungemach, 2013).

4.4.2 Ferrara (Italy): integration from different sources

In the town of Ferrara, a medium-size settlement in the north of Italy, the local heating company supplier (Energy Resource Environment Holdings (HERA)) is operating an important geothermal district heating network, which is a good example of integration of different energy sources (Figures 4.10–4.12).

The integrated energy system in Ferrara is using the following sources:

- Geothermal source
- Recovery from waste-to-energy plant (WTE)

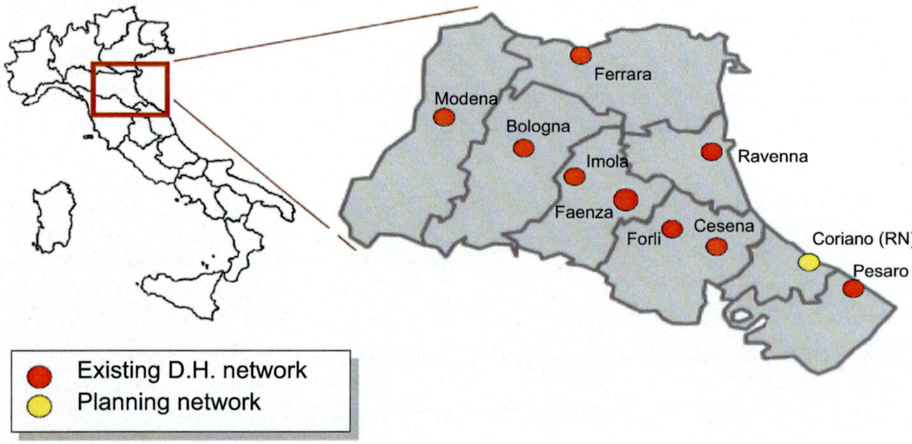

Figure 4.10 Ferrara district heating (Ferraresi, 2013).

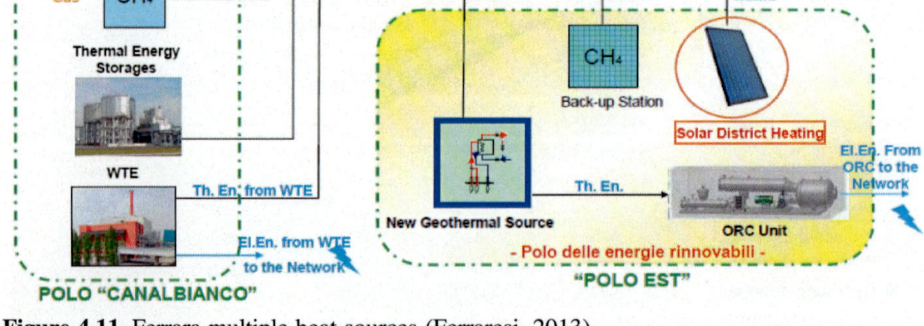

Figure 4.11 Ferrara multiple heat sources (Ferraresi, 2013).

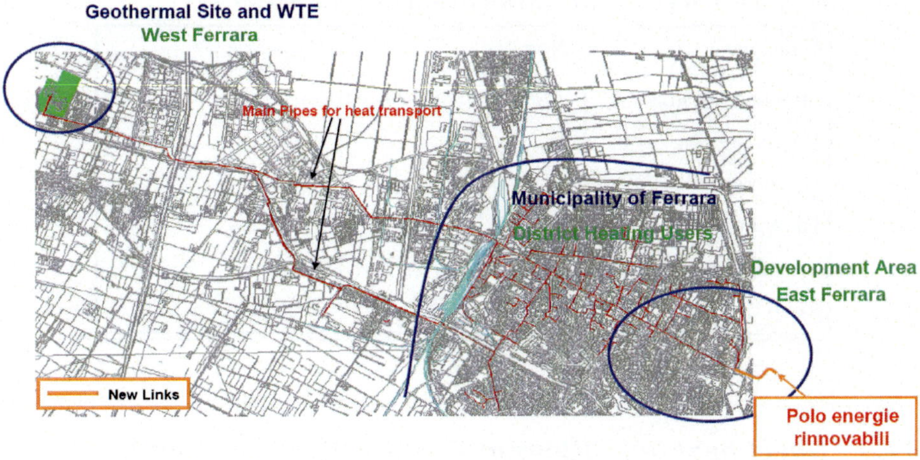

Figure 4.12 Ferrara future developments (Ferraresi, 2013).

- Backup stations
- Solar heating station (planned)
- Thermal storage
- Organic rankine cycle (ORC) electricity generation (planned)

The actual geothermal system is made from two 1-km-deep production wells, with a total flow rate of 400 m^3/h at a temperature of 100° to 105 °C; the temperature of the fluid in district heating network is from 90° to 95 °C ingoing, cooled down to 60 to 65 °C, with a thermal nominal power 14 MW_{th}, producing 75,000 MWh_{th}/year.

The WTE has an authorized capacity of waste disposal of 130,000 tons, producing 13 MW_e and 29 MW_{th}, dispatching to the heating network 80,000 MWh_{th}/year.

The backup boiler stations of 84 MW_{th} are used for covering the peak demand, in conjunction with the thermal energy storage system (approx. 1000 m^3 each one), used for reducing the usage of the boilers during the daily consumption peaks.

A typical fraction per year of the present renewable or recovery sources (geothermal and WTE) on the total thermal energy is 80%, with an energy saving 14,800 TOE and avoided emissions of:

- NO_x 47,650 kg
- SO_2 36,628 kg
- CO_2 39,411 tons

The present network will be expanded up to a total of 40,000 standard flats (40% of the town), with a solar heating station of 1 MW_{th} and a new geothermal triplet (two production and one reinjection) for 14 MW_{th}, with additional electricity production from a small low-temperature binary cycle of 1 MW_e. The final expected goal will be to reach 90% of energy from renewables, with 163 GWh_{th} geothermal, 99 GWh_{th} WTE, and 1 GWh_{th} solar.

Table 4.4 Direct heat applications in Tuscany

Category	Number	TJ/year
Individual heating	40	13.12
District heating	13	394.41
Greenhouses	6	523.39
Food processing	3	3.02
Industrial applications	3	72.75
Total	**65**	**1006.69**

4.4.3 Direct heat utilizations in Tuscany (Italy) from high enthalpy geothermal steam

In the warm heart of Tuscany, Italy, steam is producing electricity (with 34 producing units, for an installed capacity of 875 MW_e and a production of 5.2 TWh/year of electricity) and delivering a large amount of heat for several different direct utilizations: individual and district heating, greenhouses, food processing, and industrial applications, for a total dispatch of about 1000 TJ/year in the entire geothermal region (Table 4.4).

The heat is provided in different ways: low-pressure steam wells from the separated fraction of water dominated reservoirs and some from primary steam diverted from the electrical production to the heat exchanging station when required by the thermal load.

4.4.4 Example number 5: Beijing (China): cascade applications

Nangong is a normal small village in the southwestern suburbs of Beijing. In recent years, comprehensive geothermal development and usage have made great achievements. In the year 2000, the drilling of a 3000-m-deep productive well was completed, producing 2700 tons/day of water at 75 °C. This water is used in a two-stage process: first, the heat (with several cascade utilizations); second, the mineral contents for medical and gymnastic facilities. Six major projects are online: geothermal greenhouses, fish farming, fishing center, entertainment center, district heating system, and geothermal exhibition/education center. The greenhouse occupies 1.5 km^2 with 12 buildings, for flowers, vegetables, and seasonal fruits. The fish breeding and fishing area is the biggest indoor center for fishing entertainment in China, with 220,000 m^2 of covered area. The hot spring water entertainment (18,000 m^2) serves 1500 customers per day, with pool water temperature at 20 °C all year round and other heated gymnastic facilities (Figure 4.13).

The villagers of Nangong have already been benefitted by geothermal development, with all the houses geothermally heated and high-quality hot spring water available at home. A new greenhouse system for 20,000 m^2 and a new hot spring water gymnastic center of 30,000 m^2 are planned.

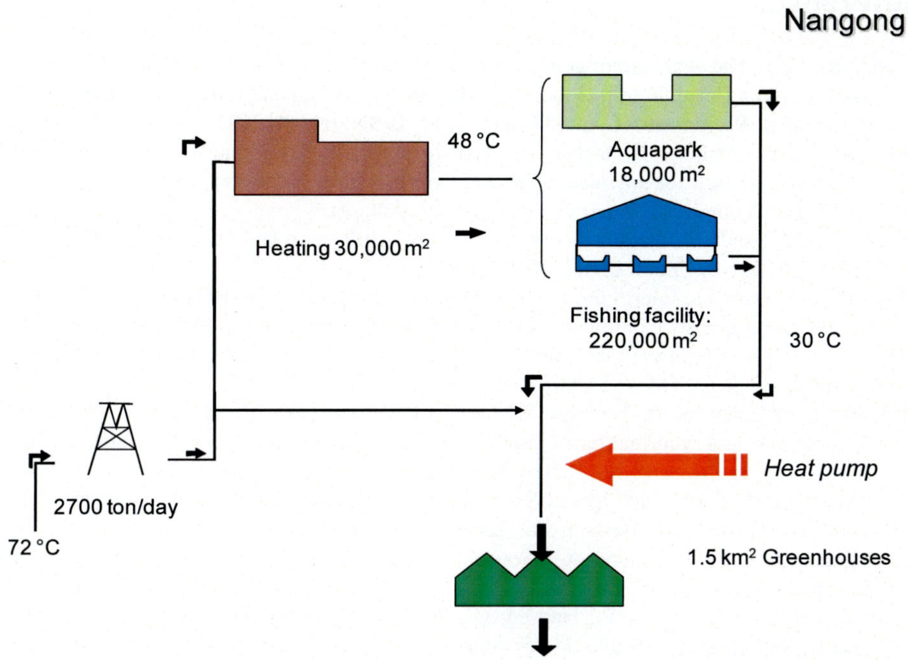

Figure 4.13 Nangong cascade applications.

4.5 Conclusion

Direct heat utilization worldwide is reported in 78 countries, with a continuous increase from the 28 identified in 1995. The last available data on thermal energy used are 438,071 TJ/year (121,696 GWh/year), with about a 60% increase over 2005, at a compound rate growing at 9.9% annually. The energy savings is impressive, about 307.8 million barrels (46.2 million tons) of equivalent oil annually, preventing 46.6 million tons of carbon and 148.2 million tons of CO_2 being released to the atmosphere. The largest annual energy use is in the following five countries: China, USA, Sweden, Turkey, and Japan (about 55% of the total).

It is very important to highlight the significant contribution to a country's or region's energy mix of the low- to moderate-temperature geothermal resources in direct heat applications. Quite often, given the right conditions, these applications can be operated in an economical approach, especially considering the constant increase in oil and gas prices. Geothermal energy for heating and cooling can be developed anywhere with geothermal heat pumps.

Low- to moderate-temperature geothermal resources are also being used in combined heat and power plants, in which hot waters are first run through a binary power plant and then cascaded for space, swimming pool, greenhouse, and aquaculture pond heating, before being injected back into the aquifer. These applications can maximize the use of the resources and improve the economics.

Geothermal energy is becoming increasingly more competitive with fossil fuels, and its environmental benefits should accelerate in the future.

References

Cataldi, R., 1999. The year zero of geothermics. In: Cataldi, R., Hodgson, S., Lund, J.W. (Eds.), Stories from a Heated Earth. Geothermal Resources Council and International Geothermal Association, Sacramento, CA, USA, pp. 7–17, ISBN: 0934412197.

Curtis, R., Lund, J., Sanner, B., Rybach, L., Hellström, G., 2005. Ground source heat pumps – geothermal energy for anyone, anywhere: current worldwide activity. In: Proceedings of the World Geothermal Congress, April 24–29, Antalya.

Dickson, M.H., Fanelli, M., 2003. Geothermal Energy: Utilization and Technology. Publication of UNESCO, New York, ISBN: 231039156.

EPRI, 1978. Geothermal Energy Prospects for the Next 50 years – ER-611-SR, Special Report for the World Energy Conference. Electric Power Research Institute, Palo Alto, CA, USA.

Ferraresi, F., 2013. International course on geothermal heat pumps. In: Short Course at EGC2013, EGEC, IGA and UGI, June 2–8, 2013, Pisa.

Fridleifsson, I.B., Bertani, R., Huenges, E., Lund, J.W., Ragnarsson, A., Rybach, L., 2008. The possible role and contribution of geothermal energy to the mitigation of climate change. IPCC Scoping Meeting on Renewable Energy Sources, Luebeck, Germany, January 21–25, 2008, p. 36. www.ipcc.ch/pdf/supporting-material/proc-renewables-lubeck.pdf.

Goldstein, B., Hiriart, G., Bertani, R., Bromley, C., Gutiérrez-Negrín, L., Huenges, E., Muraoka, H., Ragnarsson, A., Tester, J., Zui, V., 2011. Geothermal energy. In: Edenhofer, O., Pichs-Madruga, R., Sokona, Y., Seyboth, K., Matschoss, P., Kadner, S., Zwickel, T., Eickemeier, P., Hansen, G., Schlömer, S., von Stechow, C. (Eds.), IPCC Special Report on Renewable Energy Sources and Climate Change Mitigation. Cambridge University Press, Cambridge, United Kingdom, and New York, NY, USA.

Hance, C.N., 2005. Factors Affecting Costs of Geothermal Power Development. Geothermal Energy Association, for the U.S. Department of Energy, Washington, DC, USA, 64 pp.

Kagel, A., 2006. A Handbook on the Externalities, Employment, and Economics of Geothermal Energy. Geothermal Energy Association, Washington, DC, USA, 65 pp.

Lund, J.W., 2008. Utilization of Geothermal Resources. Geo-Heat Center, Oregon Institute of Technology, 2008.

Lund, J.W., Freeston, D.H., Boyd, T.L., 2010. Direct utilization of geothermal energy 2010 worldwide review. In: Proceedings World Geothermal Congress 2010, Bali, Indonesia, April 25–30, 2010.

Pollack, H.N., Hurter, S.J., Johnson, J.R., 1993. Heat flow from the Earth's interior: analysis of the global data set. Reviews of Geophysics 31 (3), 267–280.

Rafferty, K.D., Gene Culver, P.E., 1998. Heat Exchanger. Geo-Heat Center Bulletin, Klamath Falls, Oregon, March 1998, pp. 20–26. http://geoheat.oit.edu/bulletin/bull19-1/art4.pdf.

Ungemach, P., 2004. Carbonate geothermal reservoir management. In: France Int. Geoth. Days, IGD Poland 2004. Low Enthalpy Geothermal Resources Exploitation and Development, Zakopane, September 13–17, 2004.

Ungemach, P., 2013. International course on drilling, completion and testing of geothermal wells. In: Short Course at EGC2013, EGEC, IGA and UGI, June 2–8, 2013, Pisa.

Shallow geothermal and ambient heat technologies for renewable heating

5

Javier F. Urchueguia
Technical School of Industrial Engineering, Universitat Politécnica de Valencia, Valencia, Spain

5.1 Introduction—ambient heat and renewable energy

Heating and cooling are responsible for almost half of the final energy demand in Europe and are therefore the dominant ways in which energy is delivered and used. In the context of the current increasing concern about security of supply and sustainability, no serious approach to the energy problem can be made without considering heat. In this chapter, we will make an approximation to "ambient heat," understood as heat that is naturally around us in its diffuse and extended forms, originating from a diversity of heat sources, including earth, water, or air.[1]

Characteristic of ambient heat is the need of intensifying and concentrating its temperature to be able to make practical use of it, because the temperatures of the available heat sources are too low for practical heat delivery from natural temperatures. This temperature increase requires in most cases the use of some type of thermodynamic device that, by spending some amount of extra energy (from fossil sources, electricity, or even heat) is capable of delivering the heat at a useful temperature for the demand. The paramount and most utilized type of thermodynamic machine is, in this regard, the heat pump, and, thus, the concept and utilization of ambient heat is intimately connected with the possibilities offered by heat pump technologies.

Within the different types of sources of ambient heat, two are especially representative: the use of air as heat source (air source heat pumps or systems) and the use of the soil or subsurface heat (shallow geothermal systems). Both, while having the heat pump as a common element, have quite different characteristics from the point of view of design, potential, and applications. In the present chapter, special emphasis is made on the shallow geothermal technologies, less well known than air source systems.

[1] Sometimes "ambient energy scavenging," also called energy harvesting, is understood more widely in the sense of obtaining usable energy from any natural and human-made source that is in the environment, including electromagnetic, metabolic, or other sources.

Renewable Heating and Cooling. http://dx.doi.org/10.1016/B978-1-78242-213-6.00005-9

Another important point of discussion is the renewable character of the heat delivered by heat pumps, for, as pointed out before, some use of external energy is needed to ensure a given heat flow between the ambient and the demand. Traditionally, this need of an extra source (usually derived from fossil energy) has prevented most ruling bodies or institutions from considering heat pumps and, thus, the use of ambient energy as "renewable energy." But this vision has been progressively changing in accordance with a more correct understanding of the underlying thermodynamic principles, the progress in technology, and the need for better clarification of our energy sources. From the point of view of thermodynamics, part of the heat delivered by a heat pump is renewable provided that:

1. The source of ambient heat it uses is in natural equilibrium with the primary sources of renewable energy such as the sun, earth, etc.
2. The heat delivered to the demand is larger than the overall amount of non-renewable energy that was needed to ensure this flow of heat. This includes the production and grid losses, for instance, when using electricity to operate an electrically driven heat pump. Strictly speaking, only the difference of both amounts of energy can be accounted for as renewable heat production.

The above principles are embedded in the European Directive 2009/28/EC (EU Commission, 2009)—known as the Renewable Energy Sources (RES) Directive—which defines minimum efficiency standards and measurement of renewable output for heat pumps. It states that:

"The amount of aerothermal, geothermal or hydrothermal energy captured by heat pumps to be considered energy from renewable source for the purposes of this Directive, E_{res}, shall be calculated in accordance with the following formula:

$$E_{\text{res}} = Q_{\text{usable}} \times (1 - 1/\text{SPF})$$

in which:

- Q_{usable} = the estimated total usable heat delivered by heat pumps fulfilling the criteria referred to in Article 5(4), implemented as follows: Only heat pumps for with SPF $> 1.15 \times 1/\eta$ shall be taken into account
- SPF = the estimated average seasonal performance factor for those heat pumps
- η is the ratio between total gross production of electricity and the primary energy consumption for electricity production and shall be calculated as an EU average based on Eurostat data.

By January 1, 2013, the Commission will establish guidelines on how Member States are to estimate the values of Q_{usable} and Seasonal Performance Factor (SPF) for the different heat pump technologies and applications, taking into account differences in climatic conditions, especially very cold climates."

From its formulation we can see—in accordance with principle 2—that there is an intimate relationship between the amount of renewable energy and the efficiency of the heat pump or, more correctly, of the heating system as a whole. This efficiency, quantified by the SPF (which we shall more deeply discuss in the next sections), is critical to determine whether a given heat pump–driven heating system is in fact producing "renewable energy" and in what amounts.

In a strict sense, all heat sources connected to a heat pump, except waste heat, have the potential to be legally considered as renewable, depending on the system SPF they allow. Low SPF systems, however, share the risk of failing to meet the desired SPF threshold (which depends on local conditions). Consequently, to understand more correctly the shallow geothermal and other ambient heat sources in connection with the question of renewable energies, we shall discuss more thoroughly in the next chapters the most important technological aspects related to heat pumps and their efficiencies.

Finally, it is important to mention the question of cooling, which is a major requirement in modern societies and becoming increasingly more important, not only in areas where its importance is traditionally well established (so-called warm climate areas) but also in certain application areas in colder climates. From the point of view of thermodynamics, cooling is, however, a reverse flow of heat from the application to the ambient and thus cannot be considered as heat production. There are nevertheless two questions that relate the cooling issue with the topics discussed here. On one hand, the heat pump technologies related to cooling are the same as for heating. Furthermore, in some applications, the heat balance in the ambient source considered (and this is notably the case when dealing with shallow geothermal systems for heating *and* cooling) depends on the net flow of heat in either way. In this case, it is said that the source is acting as a *heat store*, in which basically an important part of the heat utilized is not coming from the natural sources of ambient heat, but is being continuously transferred to or from the application.

5.2 Technology overview

5.2.1 The basic thermodynamic principle of a heat pump and the factors affecting efficiency and SPF

5.2.1.1 Basic description of a heat pump and its components

A heat pump is a machine or device that moves heat from one location (the "source") to another location (the "heat sink") using mechanical work, according to the schematic diagram depicted in Figure 5.1.

The advantage of pumping heat is that it takes less electrical energy than it does to convert electrical energy into heat (as in electric furnaces or radiant heaters); nevertheless, it is subject to the limitations from the second law of thermodynamics, and therefore a maximum efficiency can be calculated from the Carnot cycle.[2]

[2] A European standard for testing and rating heat pump performance, EN 14511—Part 1, defines a heat pump as follows: "[a] heat pump [is an] encased assembly or assemblies designed as a unit to provide delivery of heat. It includes an electrically operated refrigeration system for heating. It can have means for cooling, circulating, cleaning, and dehumidifying the air. The cooling is by means of reversing the refrigeration cycle."

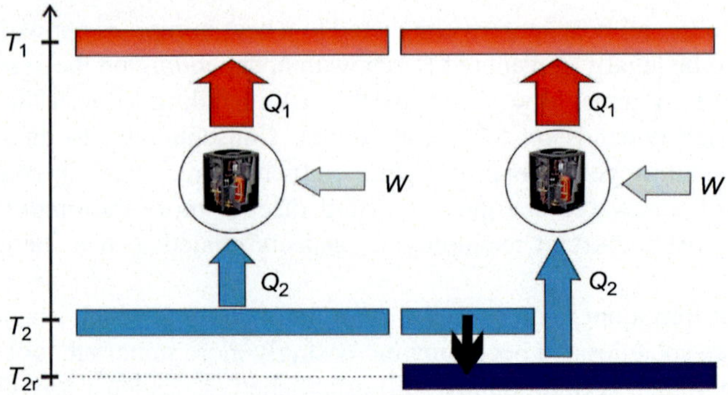

Figure 5.1 A heat pump extracts heat from the ambient at temperature T_2 and supplies heat to the demand at a (higher) temperature T_1. The mechanical work to allow this heat flow to occur (W) depends strongly on the temperature difference to be overcome. Left side: in the case of uncoupled heat sources, usually ambient air, heat extraction does not affect the source temperature. Right side: in the case of coupled heat sources, heat extraction may substantially alter source temperatures. The real/effective source temperature is then lower than the natural or undisturbed source temperature, thus penalizing efficiency. This effect is particularly present in (but not exclusive of) GSHP systems, in which, by means of design and operation, this effect is kept under control.

A heat pump uses a "working fluid," which, in its gaseous state, is pressurized and circulated through the system by a compressor. On the discharge side of the compressor, the now hot and highly pressurized gas is cooled in a heat exchanger, called a condenser, until it condenses into a high-pressure, moderate-temperature liquid. The condensed refrigerant then passes through a pressure-lowering device like an expansion valve or capillary tube. This device then passes the low-pressure (almost) liquid refrigerant to another heat exchanger, the evaporator, in which the refrigerant evaporates into a gas via heat absorption. The refrigerant then returns to the compressor and the cycle restarts.

Although in the industry, there is clear distinction—in terms of usage, and market segments—between terms such as "refrigerator/freezer," "chiller," heat pump, and air conditioners, from the thermodynamic point of view they work under the same principles and their components and technologies are also quite close. In the next paragraphs, a very brief description of the parts of a heat pump is given. A much more extensive account about heat pumps and refrigeration technologies can be found in many handbooks and sources like Granryd et al. (2003) and IEE Geotrainet Project (2011).

Refrigerant—a substance that, experiencing phase changes, circulates through the heat pump, alternately absorbing, transporting, and releasing heat. Ideally, refrigerants should be stable chemically, noncorrosive, and safe. Good refrigerants possess a boiling point somewhat below the target temperature, a high heat of vaporization, a moderate density in liquid form, a relatively high density in gaseous form, and a high critical temperature. The main refrigerants can be classified into different groups: chlorofluorocarbons (CFCs; also called "freons"), hydrochlorofluorocarbons (HCFCs), hydrofluorocarbons (HFCs), and the so-called natural refrigerants. There are critical environmental aspects to take into account with refrigerants, such as their ozone depletion potential (ODP) and global warming potential (GWP), and there have been a myriad of regulations and phase-out policies to diminish the use of high ODP and GWP refrigerants in favor of the rest.

Compressor—a machine that pressurizes the refrigerant gas, increasing its temperature. The three most representative compressor types are:

Piston compressors—a positive-displacement compressor that uses pistons driven by a crankshaft to deliver gases at high pressure. The intake gas enters the suction manifold, then flows into the compression cylinder in which it is compressed by a piston driven in a reciprocating motion via a crankshaft, and is then discharged.
Screw compressor—a type of gas compressor that uses a rotary-type positive-displacement mechanism. The rotary-screw compressor uses two meshed rotating helical rotors within a casing to smoothly force the gas into a progressively smaller gas pocket until release.
Scroll compressor—a compressor that has two spiral walls or rotary components that intermesh with each other and are used to compress a refrigerant. One of the spiral-shaped components is generally fixed, or stationary, and the other is driven in an orbiting pattern to perform its function. Scroll compressors are simpler and more efficient than piston units.

Heat exchangers—in either of two functions:

The evaporator—a coil (heat exchanger) in which the refrigerant absorbs heat from its surroundings and boils to become a low-temperature vapor.
The condenser—a coil (heat exchanger) in which the refrigerant gives off heat to its surroundings and becomes a liquid.

There are different types of heat exchangers (condenser or evaporator) in a heat pump; the most common are:

Shell-and-tube heat exchangers—fabricated with round tubes mounted in cylindrical shells with their axes coaxial with the shell axis. There are various design considerations to be taken into account such as routing of fluids (shell or tube), pressure drop, etc. Typical shell and tube heat exchangers use a "bundle" of tubes encased in a shell, in which heat energy is transferred from hot liquids or gases flowing in through the tubes to liquid or coolant that flows over and around the tubes within the shell, capturing heat energy and flowing back out.
Plate and frame heat exchangers—usually built of thin plates. The plates are either smooth or have some form of corrugation, and they are either flat or wound in an exchanger. Plate and flat-plate heat exchangers work similarly with hot and cold liquid chambers separated by metal plates.

Expansion device—a valve that releases the pressure created by the compressor, acting as a balance system to ensure the right thermodynamic conditions of the refrigerant when entering the compressor. In the reversible type of heat pumps, used to produce heating or cooling and quite common in hot climates, there is a four-way reversing valve, the function of which is to control the direction of flow of the refrigerant in the heat pump.

5.2.1.2 Efficiency of a heat pump, terms and definitions

The primary factor affecting the renewable potential of a given ambient heat source is the efficiency with which it is possible—technically and economically—to make it useful for a given application, given by the seasonal performance factor (SPF).

Usually, the SPF refers to the performance over an entire season and accounts for the energy input and heat output cumulated over that period; then the total power output is divided by the total power input, to give the SPF:

$$\text{SPF} = \frac{\text{Cumulative heat output (kWh}_{\text{th}})}{\text{Cumulative energy input (kWh)}}$$

Although SPF is a useful and meaningful indicator for performance, it is difficult to estimate and handle, because it is affected by a variety of factors that are hard to isolate or control. It is, in any case, a factor that has to be understood from the perspective of the whole heating system, and not so much related to a single element or part.

If we want to understand more deeply the connection between the different types of ambient heat sources and systems efficiency, it is quite useful to think in terms of the so-called coefficient of performance (COP) of the heat pump or heat pump system, a measure of the heat pump's efficiency seen as an isolated component or subsystem.[3] It is determined by dividing the heat power output of the heat pump by the electrical power needed to run the heat pump, at a specific temperature, which means under stable, constant, working conditions.

In this regard it is useful to refer to Figure 5.1 (left) in which the two temperature sources are represented (the heat source at temperature T_2 and the sink at temperature T_1; $T_2 < T_1$). During a certain period of time, to extract the amount of heat ΔQ_2 from the source at temperature T_2 (the relative colder ambient) to the application, we have to invest a certain amount of mechanical work, symbolized as ΔW. Hence, COP will be equal to:

$$\text{COP}_{\text{heating}} = \frac{\Delta Q_1}{\Delta W} < \frac{T_1}{T_1 - T_2} \tag{5.1}$$

In this expression all temperatures are in absolute units, and the inequality states that any possible COP is bounded by the efficiency of a Carnot cycle between sources (also called reservoirs) at the same temperatures. If ΔQ_1 is the amount of heat delivered to the hot source or reservoir during the same period, a further relationship based on the First Law of Thermodynamics states that: $\Delta Q_1 = \Delta Q_2 + \Delta W$.

In this expression, all of these terms depend on:

1. The efficiency of the heat pump (determined mainly by the quality of its components). This includes the type of refrigerant used, which substantially influences COP.
2. The temperature of the heated or cooled water/air produced by the heat pump (distribution temperature).

[3] A related concept, often used, is the *energy efficiency ratio* (EER) that measures the steady-state cooling efficiency of a heat pump. It is determined by dividing the cooling capacity of the heat pump in Btu/h by the electrical energy input in watts at a specific temperature. The higher the EER, the more efficient the unit. There is direct relationship with the COP for cooling: EER = COP × 3.412.

3. The temperature of the incoming brine (water/anti-freeze mix) from the ground loop (in the case of a shallow geothermal heat pump system) or the outdoor air (in the case of an air source heat pump).

At the same temperature difference heat pumps are more effective for heating than for cooling because the compressor's input energy is largely converted to useful heat when in heating mode and is discharged along with the moved heat via the condenser. For cooling, the condenser is normally outdoors, and the compressor's dissipated work is rejected rather than put to a useful purpose (except when the storage capacities of the source are used).

In the next sections, to analyze the quality of different available ambient sources as heat source or reservoir, we will strongly focus on the dependency of COP on the source temperatures available (item 3 in the previous list). However, COP is also strongly dependent (as points 1 and 2 state) on the technology of the heat pump, as well as on the way heat is distributed. To make it possible to compare the performance of different heat pumps, manufacturers normally publish COP figures at standard temperatures for the heated water and the brine/air, under a well-defined norm. The current European norm is called EN 1461.[4]

5.2.1.3 Types of heat pumps

Heat pump systems are available in an array of types and combinations that can suit almost any application.

For heating purposes, they can be divided into basic types, determined by the source and the destination of the heat and the medium that the heat pump uses to either absorb or reject the heat in each of these locations.

At either of the heat exchangers, the heat transfer medium can be liquid (water, or often a glycol mixture) or air; sometimes it is a combination of the two. In describing the type of heat pump, generally the heat source is provided first, followed by the destination or heat sink. The type of heat pump is, on the other hand, closely related with the type of ambient heat source that is being used in every case.

(Ambient) Air-to-air systems—Air-to-air systems use the heat energy contained in the outside air and its vapor as a source of free heat. On the demand side, heat is delivered directly by fan-assisted units to the air in the indoor space. These heat pumps may not work adequately in extremely cold temperature conditions, in which an auxiliary heating element must be used to prevent freezing, strongly penalizing heat pump performance.

Another type of air source system is the exhaust-air heat pump. This unit uses as its thermal source the stream of air being vented from the building. Because the source air is generally at the temperature of the house interior, it will not suffer the same performance reduction as an external air source heat pump. It must be stressed that the

[4] This norm will supersede the old norm EN 255. The new norm is more conservative because it takes into account not only the electricity consumption of the compressor but also that consumed by the circulation pumps and other auxiliary elements.

exhaust-air heat pump is normally only a supplement to another heating system because its source heat has had to come from elsewhere.

(Ambient) Water-to-water—Water-to-water systems operate in the same way as air-to-air systems except that the heat source is water, generally groundwater, river or pond water, or even waste heat from factory processes. The source liquid is seldom circulated directly into the heat pump in the case of lake and sea water, due to fouling of pipes, etc. Heat is then delivered to either radiators or fan-coil units within the indoor space. In an **(ambient) water-to-air** heat pump the heat source is as described before but the heat is rejected directly to the air in the indoor space. Approvals are needed for this type of installation, and restrictions exist on the type of antifreeze solution used. Similarly, in an **(ambient) air-to-water** system, heat is absorbed from the outside air and delivered to a water-based indoor system of radiators or fan coils. Heat can also be absorbed from exhaust air by this type of heat pump. This can be a means of recovering some of the heat otherwise lost from the dwelling. Blending exhaust air with ambient air will increase the source air temperature but may accelerate the buildup of frost in the evaporator.

Finally, we will deliver specific attention to the so-called **ground source heat pump (GSHP)**, which are in fact heat pumps of the water-to-water type, but in which the primary source of ambient heat is the ground beneath the surface using closed pipe loops buried horizontally in trenches or in vertical boreholes that are connected back to the evaporator. The fluid circulating in the closed loop will normally be a water/propylene glycol or acceptable equivalent antifreeze mixture. However, some direct-acting GSHPs will use refrigerant in the closed loops. Open loops may also be used to collect water from an aquifer and discharge via a separate aquifer downstream of the water table flow.

Heat may also be extracted from surface water bodies: streams, ponds, lakes, or the sea. Such systems are normally referred to as hydrothermal. Designs of any such systems must take careful account of water quality and water temperature. In all applications connected with the use of water bodies, permits are normally required from the environmental agencies.

5.2.1.4 Energy efficiency of heat pump systems: strongly and weakly coupled heat pumps

From Eqn (5.1) it is apparent that, although real COP values are usually much lower than the corresponding Carnot limit, there will be a tendency of COP to increase as the temperature difference decreases between heat source and heat sink. Hence, COP can be maximized at design depending on the system chosen. In heating mode, a system is selected requiring low temperature and a heat source with a high temperature, whereas in cooling mode we would select a system requiring high temperature and a heat source with a low temperature.

We can consider this influence more carefully if we slightly complicate the idealistic picture shown in the left-hand side of Figure 5.1 and add the temperature T_{2r} at which heat is really extracted from the heat source, in which $T_{2r} < T_2$. This accounts for the fact that in some cases heat extraction affects the temperature of the source producing a certain cool-down effect, having a potentially large influence on COP. It is therefore appropriate to talk about a coupling between the heating system and the ambient heat source. The coupling is stronger if either the thermal capacity or the heat extraction capacity of the heat exchanger in contact with the heat source is limited. In **weakly coupled systems** (usually, air source systems, or certain open-loop systems in contact with large water bodies) no account is made for the coupling effect, and thus it is usually assumed that $T_{2r} \approx T_2$. In **strongly coupled systems**—whereby shallow geothermal heat pump systems are a particularly clear example—such an approximations cannot be made and only a careful design shall ensure that heat can be extracted sustainably, without a constant degradation of system COP.

In the next section we will discuss the main factors to be considered when designing a shallow geothermal heating system. In weakly coupled systems, on the contrary, it is sufficient, based on the knowledge of T_2, to size the heat exchangers in contact with the ambient heat source based on the maximum heating capacity that the system is supposed to deliver.

Finally, one has to consider the design of the central heating system, which has a strong impact on COP through its effect on T_1, the (real) temperature at which heat is delivered to the heat sink. The highest COPs for heating are obtained in well-insulated properties with carefully designed underfloor heating.

5.2.2 Shallow geothermal systems—coupled sources

5.2.2.1 Terms and definitions

From the regulatory point of view, a clear definition for geothermal energy is stated in the EU legislative framework; EU Directive 2009/28/EC on Promotion of RES:

> *Art. 2: The following definitions also apply:*
> *(c) "geothermal energy" means energy stored in form of heat beneath the surface of solid earth.*

Contrary to deep geothermal, shallow geothermal systems can be considered as those not aiming at the higher temperatures typically found only at greater depth. Instead, shallow geothermal technology makes use of the relatively low temperatures offered in the uppermost first hundreds of meters.

The undisturbed ground temperatures that are usually found for heat extraction or injection vary between <2 and >20 °C, depending upon the climatic condition of the region and the depth of the borehole (see IEE Geotrainet Project, 2011). There are a number of different methods to transfer heat out of or into the ground. Principally, we can establish a first classification into what are called the "closed" systems, in which the water or brine inside the ground-coupled heat exchanger does not enter in direct contact with the surrounding ground, and "open" systems, methods producing water from the ground and having a heat exchanger (e.g., the evaporator) above ground. Table 5.1 summarizes some of the characteristics, typologies, and features of these systems.

Table 5.1 Classification of the most common types of shallow geothermal systems and their main characteristics

Classification	Technology	Typical depths (m)	Main advantages	Main disadvantages
Closed	Horizontal Vertical Pile or foundation heat exchange	1–3 10–250 5–50	No regular maintenance and relative environmental safety Can be used virtually everywhere[a]	Limited capacity per borehole Relatively low temperature level of heat source/high level of cold source
Open	Water wells Water from mines or tunnels	4–50	High capacity with relatively low cost Relatively high temperature level of heat source/low level of cold source	Requires maintenance of well(s) Requires aquifer with sufficient yield Water chemistry is a potential problem

[a]There are, however, areas where GSHP drilling poses a potential geological risk, like the presence of artesian waters or certain types of soil.

It is also important to mention that, besides the use of the ground as heat source by means of heat pumps, there are systems aimed at increasing/decreasing the temperature in the ground by storing heat or extracting heat (termed *Underground Thermal Energy Storage*, UTES). UTES systems make use of the same basic technologies as shallow geothermal systems and even, sometimes, as in mixed climate applications, in which GSHPs are used for heating and cooling, in principle the ground is also used as heat store.

5.2.2.2 Internal arrangements for closed shallow geothermal systems

Depending of the type of heat exchange medium inside the ground loop, there are several internal arrangements possible for closed shallow geothermal systems.

- The most common setup is the use of an intermediate fluid as heat carrier (typically water with the addition of an antifreeze agent), which is circulated through the ground loop by pumping.
- Direct expansion (DX) systems, which are not very common, are characterized by the extension of the refrigeration cycle directly into the ground loop. This avoidance of an intermediate fluid in principle improves system efficiency. However, there are drawbacks, and, in practice, DX has been applied successfully to GSHPs with horizontal loops, whereas the combination with vertical loops is problematic.
- A further possibility is the arrangement known as "heat pipes." These make use of a two-phase system inside a single, vertical pipe, in which the heat exchange fluid—a low-boiling-point substance—is evaporated by the geothermal heat in the lower section of the pipe. Steam rises then to the top of the pipe due to buoyancy forces and transfers heat to the refrigeration circuit via a heat exchanger. The steam thus cools down and condenses again, flowing back in liquid form toward the bottom of the pipe. Hence, the natural driving force is gravity, and the avoidance of pumping devices allows a potentially more efficient operation. Heat pipes are only suitable for heating and in practice not commonly used.

5.2.2.3 Heat transfer within a borehole heat exchanger and its main affecting factors

In accordance with the main principles discussed before when dealing with strongly coupled ambient heat sources, in the design of a borehole heat exchanger (BHE), it is fundamental to ensure—in a cost-effective way—that heat can be injected or extracted from the ground without excessive temperature differences between the heat carrier fluid and the surrounding ground, thus minimizing the difference between T_{2r} and T_2 (refer to Figure 5.1). This temperature difference strongly depends on a parameter known as fluid-to-ground thermal resistance, in which the two major parts of this resistance are the thermal resistance between the heat carrier fluid and the borehole wall, known as the **borehole thermal resistance**, and the thermal resistance of the surrounding ground from the borehole wall to some suitable average temperature level termed **ground thermal resistance** (see Eskilson, 1987; Hellström & Kjellsson, 1998).

Ground thermal resistance involves the surrounding ground from the borehole wall to some reference temperature level, usually the natural undisturbed ground temperature T_2 in GSHP-type applications.[5] In this type of application, it is convenient to consider the thermal response due to a step change in specific heat injection rate q (W/m)[6] given per unit length of the borehole and to associate the temperature evolution with a time-dependent ground thermal resistance R_g, so that:

$$T_b - T_2 = q\ R_g \tag{5.2}$$

where T_b is the temperature in the borehole wall. The unit of the ground thermal resistance R_g is K/(W/m). The other important factor for the design of borehole systems is the thermal resistance between the heat carrier fluid in the borehole flow channels and the borehole wall. The fluid-to-borehole wall thermal resistance gives the temperature difference between the fluid temperature in the collector (T_f) and the temperature at the borehole wall (T_b) for a certain specific heat transfer rate q (W/m):

$$T_f - T_b = q\ R_b \tag{5.3}$$

As T_f is representative of the real temperature at which the heat pump actually is taking heat from the cold reservoir ($T_f \approx T_{2r}$), from the combination of Eqn (5.2) and Eqn (5.3) we can easily deduce:

$$T_{2r} - T_2 = q\ (R_g + R_b) \tag{5.4}$$

Hence, from the perspective of system performance, we can see that it is important to minimize ground thermal as well as borehole thermal resistance. However, ground thermal resistance depends strongly on factors such as the ground heat resistivity (depending on soil type or composition) that cannot be changed by the designer. It is also important to note that usually multiple borehole arrays are used. Thermal interaction between adjacent boreholes will develop after a relatively short time, affecting the value of R_g. The usual approach here is to measure—by the so-called pulsed or transient response test (TRT) methods (ASHRAE, 2002; Gehlin, 1998)—the individual borehole value of R_g, which is then extrapolated by modeling, via appropriate so-called g-functions, to the behavior of the whole borehole field. Finally, R_g also depends on how intensively the ground was used before for thermal extraction/injection and thus on the energy behavior of the system (characterized by the number of hours the system has been used at full load throughout the heating season).

Borehole thermal resistance depends on the arrangement of the flow channels and the thermal properties of the materials involved. Typical values observed in field tests range from 0.01 K/(W/m) for the open coaxial arrangement to about 0.25 K/(W/m) for

[5] In thermal storage applications, the local average ground temperature T_m is more appropriate to characterize heat transport.

[6] Maximum thermal capacity at the heat pump evaporator (refrigeration capacity), divided by the total length of BHE, given in watt per meter BHE length (W/m).

single U-pipes in bentonite grout with poor thermal contact to the surrounding borehole wall. For a typical heat transfer rate of 50 W/m, the corresponding temperature differences that may arise due to borehole thermal resistance would range from 0.5 °C to values as high as 12.5 °C, with a potentially very significant effect on the system performance. To minimize R_b, filling materials (e.g., bentonite, concrete, etc.) in grouted boreholes are used to ensure better heat transfer. However, in water-filled boreholes—very popular in the north of Europe—the heat transfer induces natural convection in the borehole water and in the surrounding permeable ground. This effect is only possible when certain ground conditions are met and leads to a reduction of the overall borehole thermal resistance.

In general, borehole thermal resistance depends:

- On the quality of the grout
- On the borehole pipe material
- On the fluid flow inside the BHE—if flow conditions are laminar, thermal contact is much poorer than in turbulent flow conditions
- On possible thermal short-circuiting between the downward and upward legs inside the BHE

Using higher flow rates can minimize the last two factors, but there is a trade-off with the increased pumping needs that result.

Finally, a different way to go when looking at Eqn (5.4) is to keep the specific rate of heat extraction, q, limited. This is the approach of some of the best known standards for shallow geothermal development such as German VDI norm 5450 (VDI, 2008). Here, maximum tolerable heat extraction rates are fixed as a function of different soil and operation parameters of a given system. In the early years of BHE in Europe a value of 50 W/m was given as a standard value for Germany (whereas 55 W/m was the value usually adopted for Switzerland). These values were used for the design of residential GSHP at that time—and 50 W/m is still the rule-of-thumb value for the sizing of smaller installations. However, it is apparent from all considerations made before, that a BHE system should not be designed following such types of rules.

5.2.2.4 Borehole construction

Besides more rarely used arrangements, such as the coaxial types of pipes, the most usual method to achieve the heat exchange in a borehole is to insert one or more U-shaped loops of polyethylene tubing into the borehole (IEE Geotrainet Project, 2011). Single U-pipes—the dominant industry standard in past decades—are used in Northern Europe and North America, whereas double U-pipes—a simple extension of the U-pipe concept—are common in Central Europe. In Northern Europe, the boreholes are usually filled with groundwater to a few meters below the ground surface. It is common practice, and often required in many regions, to backfill the boreholes with some sealing material (usually bentonite, concrete, or even special mixtures, so-called thermally enhanced grouts).

The construction of a BHE, including the process of drilling, is nowadays a rather standardized and well-established procedure and the different materials for GSHP are readily available from a variety of manufacturers in proven quality: prefabricated BHE, grouting material, pipes, manifolds, heat pumps, etc. Shallow geothermal energy

The following three application examples have been extracted from GroundMed (2013), where complete information is available on some of the basic system parameters.

5.3.2.3 Auto showroom of VW in Bucharest

On the basis of good previous experience, operational technical and economical results, Bucharest Midocar decided to install a closed-loop, borehole-based geothermal HVAC system at a new facility in the eastern part of Bucharest. The plans for Midocar East objective included a showroom, offices, and repair workshop located on the left bank of the Dambovita River. The development was constructed with large south- and west-facing glass surfaces, its other exterior walls and construction elements were specified with an appropriate thermal resistance for the Bucharest climate. The total thermal energy demand of the building was calculated to be 1071 MW_{th}/yr.

Country	Romania
Building type	Public/offices
Year of construction	2003
Heated/cooled building area	250 m^2
Specific heat load	72 W/m^2
Specific cooling load	64 W/m^2
Type of heat pump system	10 Electric water/water heat pumps located in 2 geothermal plant rooms
System purpose	Heating and cooling and hot water production
Year of heat pump	2011
Heating capacity	390 kW_{th}
Cooling capacity	421 kW_{th}
Specific features	Several heat pumps work continuously in cooling mode
Distribution system	Fan coil units radiant floor and hot water tank
Design heating temperature	40 °C/return: 35 °C
Design cooling temperature	10 °C/return: 15 °C
Refrigerant	n/a
BHE field description	16 × 7 array comprising 112 boreholes of 140 mm diameter with 72 m thermally active depth, BHE spacing was 5 m. The boreholes were installed with single U 33.4 mm OD PE pipe (1 inch SDR11), with shank spacers placed at 3 m intervals. The boreholes were sealed with a thermally enhanced bentonite/silica sand grout with a minimum conductivity 1.7 W/m/K.

5.3.2.4 Example of a system in a public building used for mixed application (heating and cooling) in which information exists on the comparative performance with respect to an ASHP

The system is located in an institutional building at the Universitat Politècnica de València in Spain and was the result of two EU projects, GEOCOOL and GROUNDMED. The air conditioned area comprises approximately 250 m^2 and includes corridors, offices, computer rooms, and other facilities. A parallel air source system was installed in the ceiling of the building to allow comparative SPF analysis of both systems (see Urchueguía et al., 2008). The heating/cooling distribution system in the building consists of a series of 12 parallel connected fan coil units. The existing heat pump, based on propane as natural refrigerant and from the GEOCOOL EU project, was replaced by a new one manufactured by HIREF and designed specifically for the GROUNDMED project. The system is characterized by the use of two tandem compressors of the same capacity able to follow the thermal load of the building and improve the system energy efficiency. Internal and external circulation pumps are inverter driven.

Country	Spain
Building type	Public/educational
Year of construction	2003
Heated/cooled building area	250 m^2
Specific heat load	72 W/m^2
Specific cooling load	64 W/m^2
Type of heat pump	Electric/Water/Water
System purpose	Heating and cooling
Year of heat pump	2011
Heating capacity	17 kW_{th}
Cooling capacity	15 kW_{th}
Specific features	Tandem compressors to adapt to load conditions
Distribution system	Fan coil units
Design heating temperature	40 °C/return: 35 °C
Design cooling temperature	10 °C/return: 15 °C
Refrigerant	R410A
BHE field description	6 vertical BHE of 50 m depth each (in a 2 × 3 rectangular grid)—different type of grouting—water as heat transfer fluid

5.3.2.5 Example of a system in a public building used for mixed application (heating and cooling) with free cooling capability and in build storage capacity—CIAT subsidiary building, Septèmes Les Vallons

An existing office building used as sales office and for customer service was retrofitted in 2010. It comprises 10 offices, one meeting room, one workshop, and one social room. A GSHP with six vertical BHE (100 m each), installed in 2011, provides heating and cooling (heating capacity: 26 kW, cooling capacity: 26 kW). It does not provide domestic hot water. Geocooling is implemented in the project. A PCM tank storage connected to a second geothermal heat pump is used to supply cold water to the dehumidification coil of the air handling unit. The new system replaces a 20-year-old air-to-water heat pump (30 kW) and the complete distribution system. The new distribution system consists of Coanda effect fan coil units with EC motors, one air handling unit for fresh air pretreatment with variable speed fan, as well as air heaters in the workshop and machinery room. All of the installation is controlled by a building management system.

Country	France
Building type	Public/office building
Year of construction	1990/retrof. in 2010
Heated/cooled building area	330 m^2
Specific heat load	79 W/m^2
Specific cooling load	45 W/m^2
Type of heat pump	Electric/Water/Water
System purpose	Heating and cooling
Year of heat pump	2011
Heating capacity	26 kW_{th}
Cooling capacity	26 kW_{th}
Other spec. Features	PCM tank to store cold and supply cold water to the air handling units
Distribution system	Fan coil units
Design heating temperature	40 °C/return: 35 °C
Design cooling temperature	10 °C/return: 15 °C
Refrigerant	R410A
BHE field description	6 vertical BHE of 50 m depth each (in a 2 × 3 rectangular grid)—different type of grouting—water as heat transfer fluid

5.4 Heat pump and shallow geothermal systems market overview—potential and trends[7]

5.4.1 Sources of information

Information on the worldwide development of heat pump markets and their different applications is rather heterogeneous as statistics often only count the number of units sold and not the installed capacity and utilization efficiency. The amount of renewable heat produced, energy demand, and CO_2 emission reduced by the existing heat pump stock is even less precisely known.

Europe is an exception to this rule, with the European Heat Pump Association (see its annual outlook, EHPA, 2013) and Eurostat (see EuroObserver (2013) and EuroObserver (2014)) publishing annual statistics with a varying degree of detail comparing all acknowledged RES sources and technologies including heat pumps. After the publication of the commission decision on how to calculate the renewable share from heat pumps (31.3.2013), both organizations use the official commission method for calculation. Data are collected separately for the different energy sources used by heat pumps: air, water, and ground (geothermal), and by energy distribution system (air or hydraulic). This allows for a detailed analysis of trends and developments.

Reversible air—air units used for cooling only are excluded from the RES and EE savings calculations but are reported as part of the technology's sales base in the European Heat Pump Association (EHPA) statistics.

Main sources used here are:

1. The annual heat pump market and statistics report published by the EHPA (European Heat Pump Market and Statistics, 2013).
2. The regular updates on shallow geothermal heat pumps published within the European Geothermal Congress (Antics et al., 2013) and every 5 years within the World Geothermal Congress (Lund & Boyd, 2015). These publications deal particularly with GSHPs and allow a quite complete view on worldwide capacity buildup and RES production trends.
3. The publications of the Department of Energy, particularly Goetzler (2009) for the market in the United States and for other regions such as China and Canada national/regional reports on GSHP use by Zheng (2010), Michel, Tanguay, and Grasby (2015), or Sivasakthivel and Murugesan (2014).

5.4.1.1 Early market developments

The earliest example for GSHP in the literature dates from 1945 in Indianapolis, IN, USA, describing a DX system based on horizontal loops (Crandall, 1946) and a similar vertical loop configuration as usual today (Kemler, 1947). In Europe (Austria, Germany, Sweden, Switzerland), the first GSHP with groundwater wells and the first horizontal loops appeared around 1970, and the first BHE before 1980. The second oil price crisis, around 1980, triggered a short boom followed by a slower market

[7] The work for this section was also contributed by Dr. Thomas Nowak.

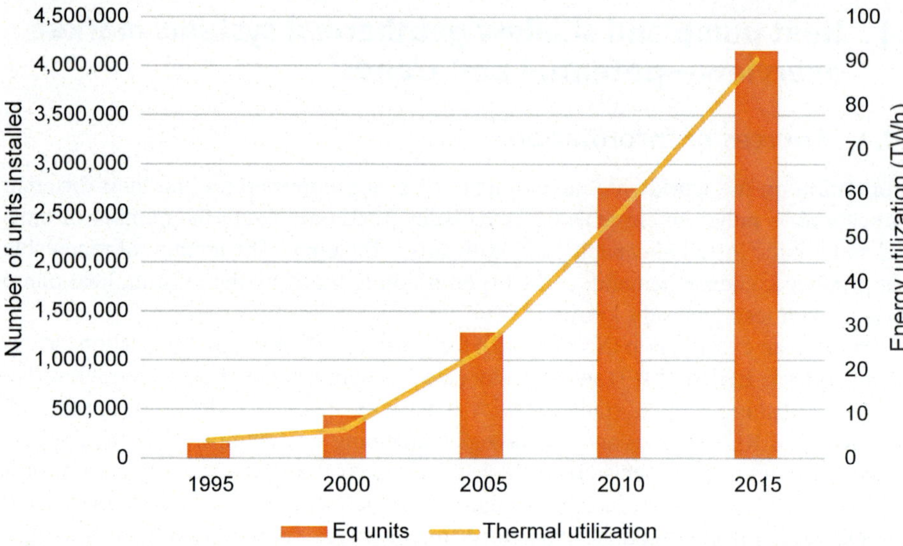

Figure 5.2 Worldwide trend of GSHP installed and energy produced between 1995 and 2015. The number of units is calculated from the installed capacity considering an average unit capacity of 12 kW_{th}. *(After Lund and Boyd (2015).)*

development in the following years, except in Sweden and Switzerland. This was also the case in the United States and Canada, where a robust GSHP market surged under the umbrella of the concept "geoexchange systems." In parallel, GSHP technology has spread gradually to all EU countries and since 2008 even to milder climate zones where reversible heating and cooling systems are mostly demanded (mainly France, Italy, but also in Spain). In Figure 5.2, the worldwide trend in installed units between 1995 and 2010 is shown in Lund and Boyd (2015), highlighting an accelerated uptake of GSHP since 1995, and especially after 2000, due to a combination of factors such as high oil prices, improved technologies and favorable incentive schemes in several important countries and regions (USA, Europe).

5.4.2 Recent developments in the European market

The European Heat Pump market has seen a strong development until 2008, followed by a phase of stagnation (2009–2013) and a return to growth (2014), which is expected to continue. In terms of energy source used, air-source-based applications dominate the market, whereas the share of geothermal heat pumps dropped to 13% in 2014.

Market development is heterogeneous across the different member states of the European Union as a result of different climate conditions, building codes, and building tradition as well as legislation, support schemes, electricity pricing, and even established market structures in the energy and HVAC sectors (i.e., how well natural gas infrastructure is established in each region).

Three main trends can be identified, that are observable across all European markets:

1. Air is, and presumably will remain, the dominant energy source for heat pumps even for heating purposes.
2. Sanitary hot water heat pumps are the fastest growing heat pump segment across Europe.
3. Larger heat pumps for commercial, industrial, and district heating applications are increasingly popular. Here, geothermal or hydrothermal energy is dominating and growth can be observed.

5.4.2.1 Heat produced by heat pumps—RES heat production and 2020 targets

Counting heat pumps as energy producers rather than mere consumers of electricity of other sources of auxiliary energy is a relatively new development. In Europe it was established for all heat pumps under the Directive on the promotion of the use of energy from renewable sources (EU Commission, 2009). The Directive set mandatory targets for the share of RES on total final energy demand, to be achieved by 2020. Member states had to present plans and roadmaps for the different technologies to be used to achieve the respective national target. The European Commission was then given the mandate to present a method on how to determine the RES share from heat pumps, which was published in 2013. Since then, heat pump contribution to RES is included in an increasing number of national statistics and subsequently on the Eurostat Website (http://ec.europa.eu/eurostat/web/energy/data/shares). It is worth noting that a minimum efficiency criterion was also established to avoid counting heat pumps with very low efficiency (see Figure 5.3).[8] It is interesting to see that there are a few countries—namely, France, Sweden, Italy, and Germany—where the heat pump RES production is significantly higher than in the rest of EU countries. It is, however, alarming that some countries that reportedly have thriving heat pump markets have chosen not to report the RES contribution. It is expected that a more mature and uniform data collection process across Europe will remedy the current situation in the near future.

It is also useful to compare RES heat produced by heat pumps with other RES and non-RES heat sources to understand the relative importance of heat pumps in energy production in the different markets. To this purpose we have collected data of the different National Renewable Energy Action Plans issued by the different EU countries for 2010 and projected for 2020 quantifying total energy consumption for heating and cooling and total RES heating energy production. The 2010−2020 trend was interpolated to our reference year 2013 to compare with data shown in Figure 5.3. One result is shown in Figure 5.4, in which the relative weights of the non-RES heating and cooling (H&C) versus RES heating is depicted for the different EU countries. It

[8] Although in global terms Eurobserver and EHPA statistics agree that the country-by-country distribution shows substantial variations. In the case of Eurobserver data, Italian figures produce a certain distortion, whereas EHPA figures are more coherent with overall heat pump figures and thus used EHPA data.

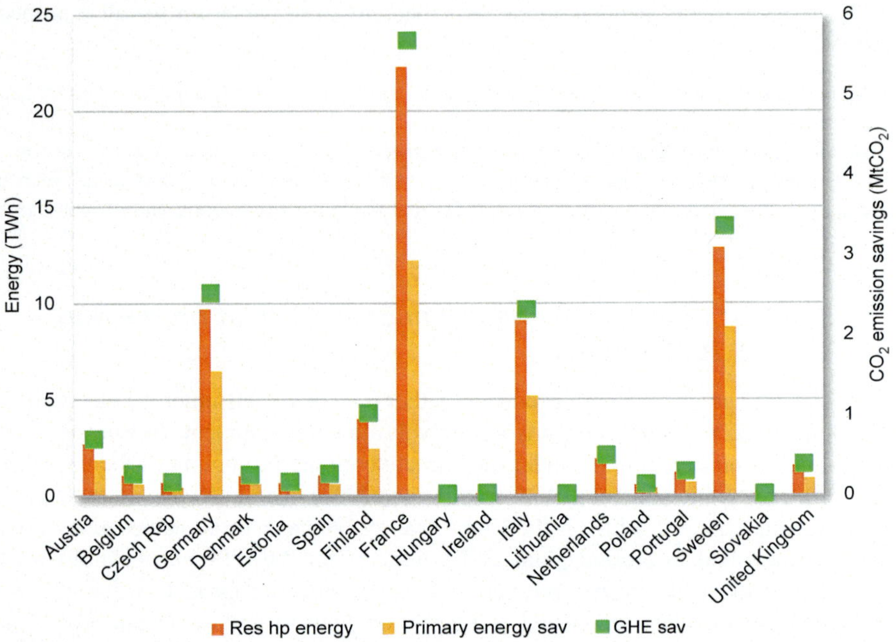

Figure 5.3 Energy and CO_2 emission savings produced by heat pumps across different EU countries according to EHPA statistics (European Heat Pump Market and Statistics, 2013). The orange and yellow bars indicate, respectively, RES energy and energy savings in TWh (left vertical axis); the green line (right axis) shows CO_2 savings in MCO_2 t/year.

becomes obvious that the different member states assess the potential of renewable heating in general and heat pump technology in particular very differently. Some countries like Sweden report more than 60% of heating from RES sources (dominated by biomass) and others, like the UK, report less than 5%. Furthermore, in Figure 5.5, the different countries are represented according to their RES/non-RES H&C ratio and HP RES/non-HP RES heating production to focus more on the contribution of heat pumps. Clear differences can be observed between high RES H&C and high heat pump share countries (Sweden, France, Estonia), high RES H&C but low heat pump share countries (e.g., Lithuania), high-heat pump share but low-RES H&C countries (like notably Italy and Netherlands) and finally countries (like Spain, Ireland, etc.) in which RES H&C as well as heat pump share is below the EU average.

Because the 2020 target for renewable energy is not automatically reached for all countries, it might be interesting to analyze the success factors for a larger heat pump use and lobby for their adoption in countries with low heat pump use and a risk of not achieving the targets.

The generally very low ambition toward the use of heat pumps is not recognizing the proven potential. A stronger technology pickup would thus enable much more ambitious RES targets across the EU member states and worldwide.

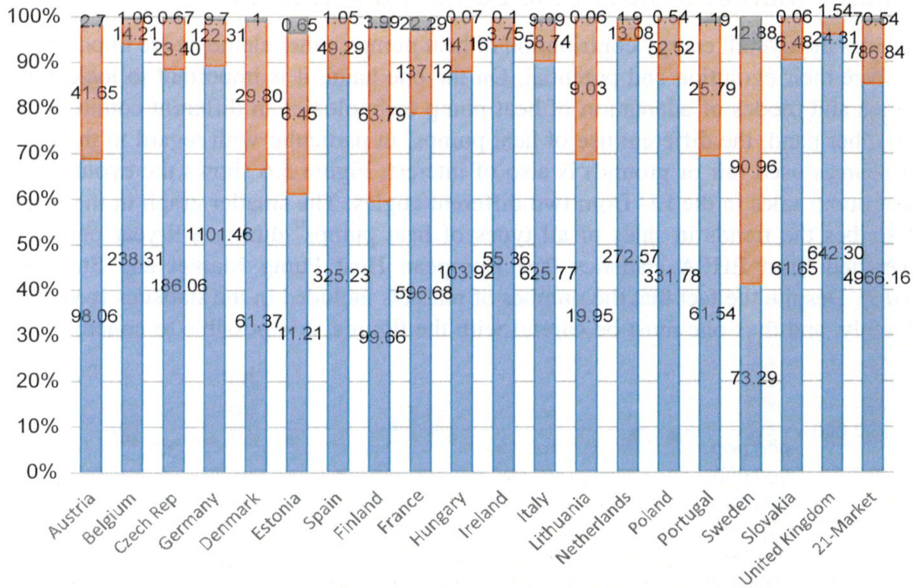

Figure 5.4 Share of energy to satisfy heating and cooling demands in the different EU countries divided in to three categories: the blue bars correspond to fossil, nonrenewable sources, the orange bars represent the contribution of renewable sources of heat (excluding heat pumps), whereas gray bars represent the heat contribution of the installed heat pumps in accordance with in EHPA statistics. Data are interpolated for year 2013 following the trend between the declared 2010 data and 2020 projections that each country has made in its respective National Renewable Energy Action Plans.

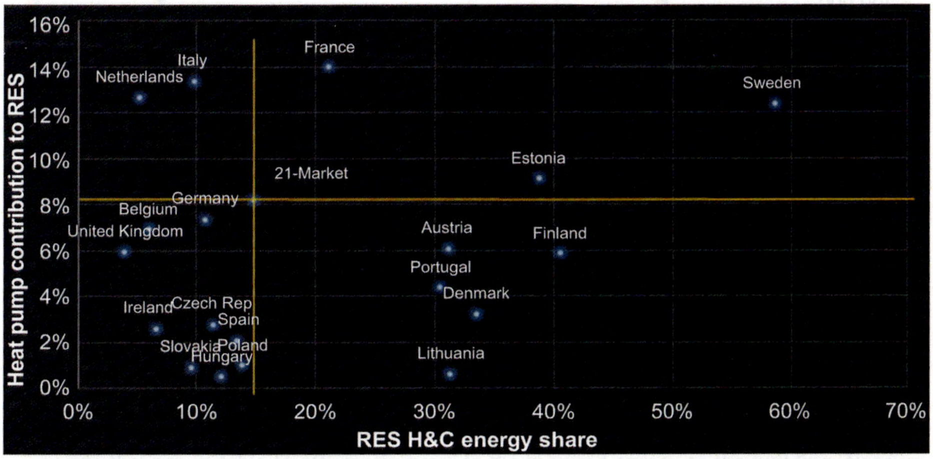

Figure 5.5 Relative contribution of RES heat to total H&C versus relative contribution to heat pumps to RES heat for the different EU countries according to NREAP provisions interpolated to year 2013. The yellow lines correspond to the EU average on the 21 markets.

3. Development of a heat pump for near-zero energy buildings (single family house)—COP (for heating and cooling) increased from 3.5 in 2012 to 6 in 2025. Contribution to the production of DHW higher than 40% in 2025.
4. High-capacity heat pump for simultaneous production of cold and hot water for heating/cooling the building. sCOP of air-to-air heat pump increased from 7 in 2012 to 10 in 2020. Refrigerant charge lower than 0.1 kg/kW.
5. Sorption cooling systems driven by hot water at moderate temperature—Driving temperature for absorption today at 95 °C diminished to 60 °C. Reference sCOP (water cooled) increased from 0.5 in 2012 to 0.8 in 2020.
6. Enhanced industrial compression heat pumps. Carnot efficiency increased from 0.3 in 2012 to 0.4 in 2025. Production cost of the heat pump unit reduced from 300 €/kW in 2012 to less than 200 €/kW in 2025.
7. Process integration, optimization, and control of industrial heat pumps. Demonstration 2014–2016. Reference sCOP compression heat pump (at $DT = 35$ K, $T_{evap} = 40$ °C) increased from 3.5 in 2012 to 5 in 2020. Reference sCOP absorption heat pump increased from 1.1 in 2012 to 1.5 in 2020. Average system cost reduced from 500 to 600 €/kW in 2012 to less than 400.

On the other hand, shallow geothermal technologies are represented by the Geothermal Panel within the Technology Platform, and the areas of research included in their roadmap include:

1. Improved vertical borehole drilling technologies to enhance safety and reduce cost of BHE installations—Improved installation technologies and geometries for ground Heat Exchange technology.
2. European-wide Geoactive Structures Alliance—Development of a network of laboratories to create four testing sites.
3. Improved pipe materials for BHE and horizontal ground loops—New pipes for higher temperatures. Improved thermal transfer fluids.
4. Creation of a new European-wide database to map conductivities and potential (to 100 m depth) and feasibility of vertical BHE systems.
5. Development of a geophysical tools for shallow reservoir potential estimation—Enhanced TRT methods for nonconventional systems.
6. Integration of design of the shallow geothermal system and building energy system with regard to optimum thermal use and operational strategy.
7. System concepts and applications for geothermal large-scale and medium-scale cooling in warm climates—Hybrid systems, new high-temperature pipe materials and new short-term storage materials and concepts. Campaign to support 50 demonstration plants.
8. Development of ground coupling technologies and installation techniques for high capacities through hybrid systems and integration with other RES sources—Campaign to support 50 demonstration plants.
9. Nontechnical provisions—Measures to increase awareness, harmonization of shallow geo-standards, shallow geothermal installer EU-wide training certificate, and shallow geothermal Smart City deployment policy along the line of previous projects.

Although the proposed research areas will contribute to enlarge the potential of heat pumps, including shallow geothermal systems, their uptake in the European Horizon 2020 program is slow. The heating and cooling sector, which historically has seen a small share of public funding for research and development compared to other energy sectors, needs continuous and substantial resources to attract talent and to eventually produce the desired results.

In the case of renewable heating, which according to the European Commission addresses one of the sectors most important for a stable, safe, affordable—in other words: sustainable—energy future, it should be easy to justify preferred support.

References

Antics, M., Sanner, B., & Bertani, R. (2013). Summary of EGC 2013 country update reports. In *European Geothermal Congress, Pisa.*

ASHRAE. (2002). *Methods for determining soil and rock formation thermal properties from field tests. ASHRAE Research Summary – ASHRAE 1118-TRP*. American Society of Heating, Refrigerating and Air-Conditioning Engineers.

Crandall, A. (1946). House heating with earth heat pump. *Electrical World, 126/19*, 94–95.

EGEC. (2015). European Geothermal Council Website.

EHPA. (2013). European Heat Pump Market Statistics Report. European Heat Pump Association.

Eskilson, P. (1987). Thermal analyses of heat extraction boreholes (Ph.D. thesis). Lund, Sweden: Department of Mathematical Physics, Lund Institute of Technology.

EU Commission. (2009). *The "Directive 2009/28/EC of the European Parliament and of the Council."*

EuroObserver. (2014). *The state of Renewable Energies in Europe - Edition 2014-14th EurObserv'ER Report.*

EuroObserver (FR). (2013). *European heat pump market and statistics*. European Heat Pump Association.

Gehlin, S. (1998). Thermal response test – in-situ measurements of thermal properties in hard rock (Licentiate thesis, 1998:37). Luleå, Sweden: Luleå University of Technology.

Granryd, E., Ekroth, I., Lundqvist, P., Melinder, A., Palm, B., & Rohlin, P. (2003). *Refrigeration engineering*. Stockholm: Department of Energy Technology, Royal Institute of Technology, KTH.

GroundMed. (2013). GroundMed project Website. Retrieved from http://www.groundmed.eu/.

Hellström, G., & Kjellsson, E. (1998). Thermal performance of borehole heat exchangers. In *Proc. of Conf. at the Richard Stockton College, Pomona, USA.*

IEE Geotrainet Project. (2011). Geotrainet training manual for designers of shallow geothermal systems. In *GEOTRAINET, EFG, Brussels, 2011.*

Jones, W. (2001). *Air conditioning engineering* (5th ed.). Butterworth-Heinemann.

Kemler, E. (1947). Methods of earth heat recovery for the heat pump. *Heating and Ventilating*, 69–72.

Lund, J., & Boyd, T. (2015). *Direct utilization of geothermal energy 2015 worldwide review*. Melbourne, Australia.

Michel, M., Tanguay, D., & Grasby, S. (2015). Direct utilization of geothermal energy from coast to coast: a review of current. In *Proceedings of the World Geothermal Congress - preprint. Melbourne, Australia.*

Sivasakthivel, T., & Murugesan, K. (2014). A study on energy and CO_2 saving potential of ground source heat pump system in India. *Renewable and Sustainable Energy Reviews, 32*, 278–293.

Urchueguía, J., Stryi-Hipp, G., Alakangas, E., Cabeza, L., Gramelis, P., Haslinger, W., & van Helden, W. (2014). *Common implementation roadmap for renewable heating and cooling*. Brussels: European Technology Platform on Renewable Heating and Cooling.

Urchueguía, J., Zacarés, M., Corberán, J., Montero, Á., Martos, J., & Witte, H. (2008). Comparison between the energy performance of a ground coupled water to water heat pump system and an air to water heat pump system for heating and cooling in typical conditions of the European Mediterranean coast. *Energy Conversion and Management, 49*(10), 2917–2923.

VDI. (2008). *VDI 4640. Thermische Nutzung des Untergrunds.Grundlagen, Genehmigungen, Umweltaspekte*. Verein Deutscher Ingenieure.

William Goetzler, R. Z. (2009). *Ground-source heat pumps: Overview of market status, barriers to adoption, and options for overcoming barriers*. U.S. Department of Energy.

Zheng, K. (2010). Growth of the use of geothermal heat pumps in China. In *Proceedings World Geothermal Congress 2010, Bali.*

Zheng, K., Mo, Y., & Chen, L. (2015). Twenty years of geothermal heat pumps in China. In *Proceedings World Geothermal Congress, Bali.*

Zheng, X. X. (2010). Geothermal direct use and its contribution to CO_2 emission saving in China. In *Proceedings of the 2010 World Geothermal Congress* (paper No. 2809, 6 pp.). Bali, Indonesia.

Solar cooling technologies

6

Uli Jakob [1,2]
[1]Dr Jakob Energy Research GmbH & Co. KG, Weinstadt, Germany; [2]Solem Consulting, Berlin, Germany

6.1 Introduction

The usual split units have maximum energy consumption in peak-load periods during summertime. In the past few years this regularly led to grids working to maximum capacity and blackouts in sunny and emerging countries. An attractive alternative to conventional electricity-driven cooling systems is solar cooling, which combines thermally driven heat pumps/sorption chillers with solar thermal collectors. In the case of solar cooling, the main advantage is the coincidence of solar global irradiation and cooling demand, which matches very well the sunny and hot climates all over the world (Figure 6.1). However, for thermal cooling, e.g., by waste heat from a biogas-driven Combined Heat and Power (CHP) unit the benefit is the longer operating time of the CHP unit itself and with that the increased electricity production.

The world's first solar cooling system was running in Paris, France (Figure 6.2), during the world exhibition of 1878 (Mouchot, 1987). This system consisted of an ammonia/water absorption chiller and a parabolic reflector to produce ice. The first commercial solar cooling systems for air conditioning were developed in Europe and the USA 100 years later, e.g., by the companies Dornier–Prinz Solartechnik, Germany (Schubert and Dreyer, 1977) and Arkla Industries, Inc., USA, and Robur, Italy (Grossmann, 2002). These systems have been realized in several demonstration projects. Because of the lack of demand on the market for solar cooling in the 1970s and 1980s, the production of these solar cooling systems was stopped.

6.2 Current market and potential

Solar cooling is especially appealing if the solar thermal system is also used for other applications such as heating, domestic hot water, etc. Thus, maximum operation time and low-cost driving heat for sorption chillers are key for economic efficiency of solar cooling systems. Several small- and medium-scale sorption chillers have been developed in the past few years, some of them especially for solar cooling applications. The global solar cooling market grew at an average annual rate exceeding 40–70% between 2004 and 2014, ending the period with about 1200 systems of all technology types and sizes—mostly in Europe (Figure 6.3). Large-scale systems are creating interest due to their more favorable economics, whereas the availability of small ($<20\ kW_r$) cooling kits for residential use has increased interest in the

Renewable Heating and Cooling. http://dx.doi.org/10.1016/B978-1-78242-213-6.00006-0

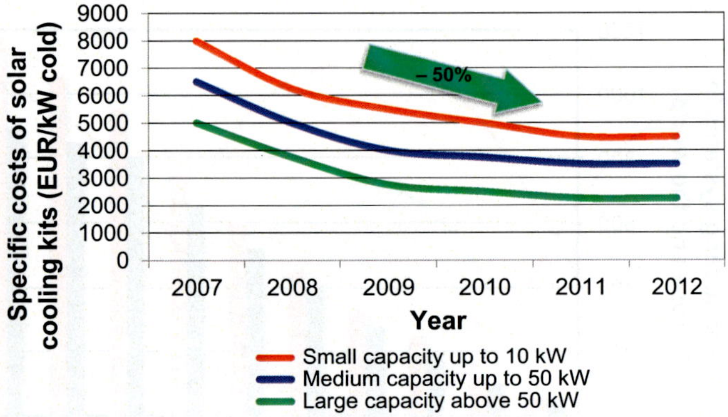

Figure 6.4 Cost development of small- to large-scale solar cooling kits (2007–2012).
Source: Solem Consulting.

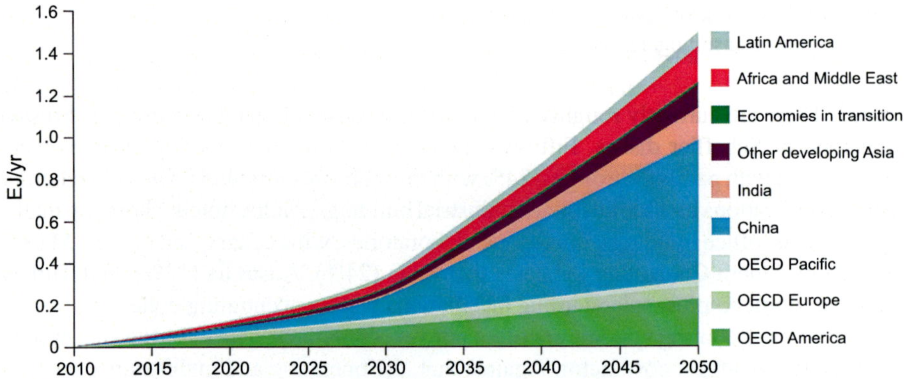

Figure 6.5 International Energy Agency Technology Roadmap solar heating and cooling—Market potential by 2050.
Source: IEA Technology Roadmap Solar Heating and Cooling, 2012.

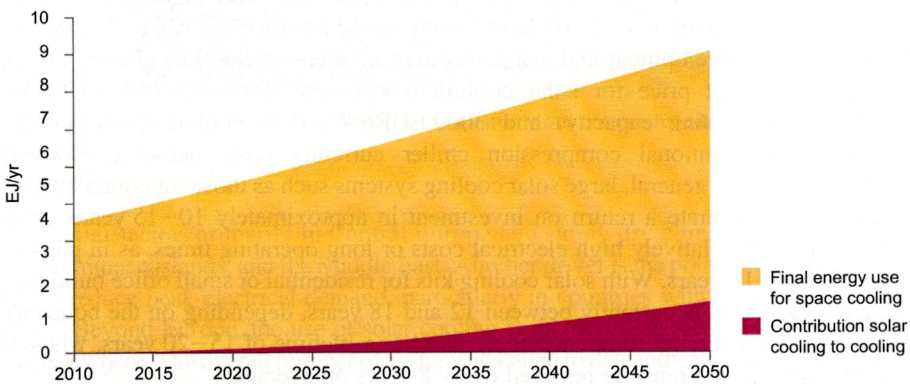

Figure 6.6 International Energy Agency Technology Roadmap solar heating and cooling—Share of solar cooling by 2050.
Source: IEA Technology Roadmap Solar Heating and Cooling, 2012.

2020 (IEA, 2012). To develop this market, different national instruments like incentive schemes or tax benefits have to be developed and introduced. One example is the German market incentive program (MAP), in which the German Federal Office for Economic Affairs and Export Control (BAFA) offers today (starting 01.04.2015) 200 EUR/m^2 for collector areas between 20 and 100 m^2 for solar thermal cooling at existing buildings or 150 EUR/m^2 now also for new buildings! Furthermore, the German BAFA offers also a solar performance-based subsidy for solar cooling of 0.45 EUR per kWh/a/collector (Solar Keymark is mandatory) for collector areas between 20 and 100 m^2 (also starting from 01.04.2015). In addition to the pure solar cooling funding, BAFA has opened the program for promotion of efficient cooling systems in industry for sorption technology/solar cooling between 5 and 500 kW$_r$ cooling capacity, which offers 25% of the repayment bonus of the net investment cost for sorption chiller systems including heat rejection (since 01.01.2014).

6.3 Solar cooling technologies for different types of solar cooling

The solar cooling technology has been used for several years, but few solar cooling systems are yet commercially available. Different key technologies like solar thermal collectors, sorption cooling technologies, and heat rejection systems (e.g., wet cooling tower, dry cooler, or hybrid cooler) are required to set up a solar cooling system depending on the application in residential buildings, commercial buildings, and industry (Figure 6.7). Therefore, Table 6.1 shows the solar thermal collector technologies and their applications for different cooling purposes. Additional the used heat transfer mediums and maximum reachable collector temperatures are given to

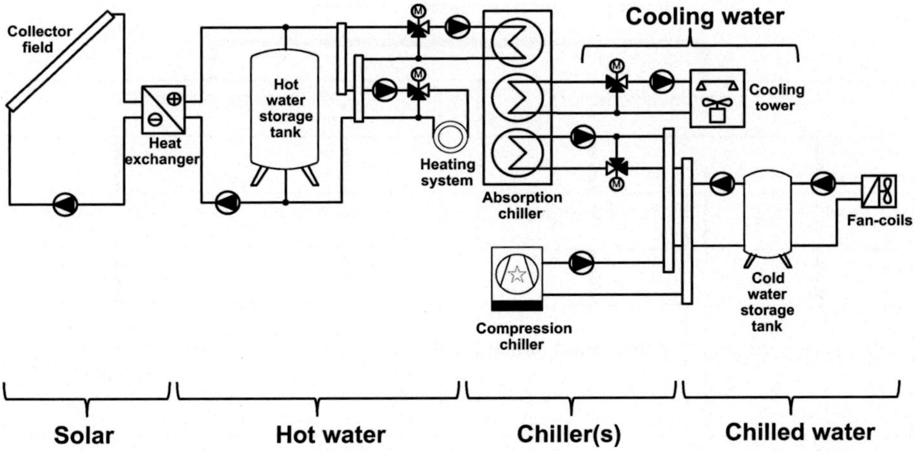

Figure 6.7 General scheme of a solar cooling system.
Source: Solem Consulting.

Table 6.1 Solar thermal collector technologies—applications for solar cooling

Solar thermal collector	Heat transfer medium	Collector temperature	Application for cooling
Air collector	Air	40–60 °C	Air conditioning
Flat-plate collector	Water, water–glycol	70–90 °C	Air conditioning, slab cooling
Evacuated-tube collector	Water, water–glycol	90–120 °C	Air conditioning, slab cooling
Parabolic trough/ Fresnel collector	Thermal oil, water	120–250 °C	Refrigeration, air conditioning, slab cooling

investigate the various solar cooling systems. The technology of thermally driven solar cooling systems is mainly divided into closed systems to produce cold water and open systems for air conditioning (Figure 6.8). Table 6.2 shows the working pairs and the cooling medium as well as the different temperature ranges for cooling, heating, heat rejection, the cooling capacity per unit, and the coefficient of performance (COP) of the presented marketable sorption cooling technologies. In principle, absorption chillers, adsorption chillers, and desiccant and evaporative cooling (DEC) systems are used for the air conditioning of buildings (Figure 6.9). However, liquid sorption systems are mainly in the prototype or field test stage. Figure 6.10 shows

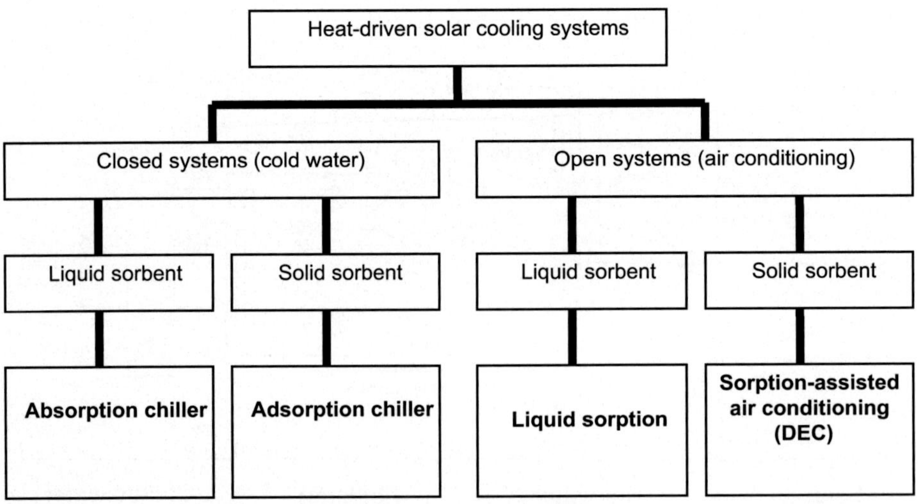

Figure 6.8 Principal classification of solar cooling systems.
Source: Dr Jakob Energy Research/Solem Consulting.

Table 6.2 Solar thermal driven or assisted cooling and air-conditioning technologies

Technology	Absorption (single-effect)		Adsorption	Liquid sorption	Desiccant and evaporative cooling
Refrigerant	Water	Ammonia	Water	–	–
Sorbent	Lithium bromide	Water	Silica gel or zeolite	Lithium or calcium chloride	Silica gel or lithium chloride
Cooling medium	Water	Water glycol	Water	Air	Air
Evaporator temperature	6–20 °C	−30 to +20 °C	6–20 °C	16–20 °C	16–20 °C
Heating temperature	60–95 °C	65–150 °C	50–95 °C	45–70 °C	55–100 °C
Heat rejection temperature	25–35 °C	25–50 °C	25–35 °C	Not required	Not required
Cooling capacity range (per unit)	10–20,500 kW_r	19–1000 kW_r	5–430 kW_r	10–350 kW_r	6–300 kW_r
Coefficient of performance (COP)	0.6–0.8	0.5–0.7	0.6–0.7	0.5–1.0	0.5–1.0

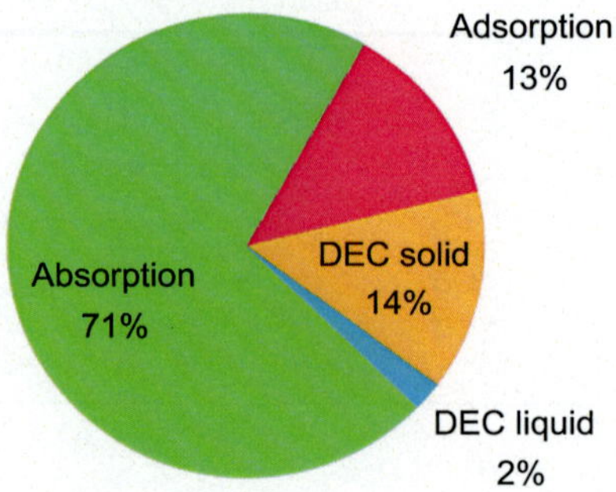

Figure 6.9 Market share of solar-driven sorption chillers (2009). DEC, desiccant and evaporative cooling.
Source: EURAC/Tecsol.

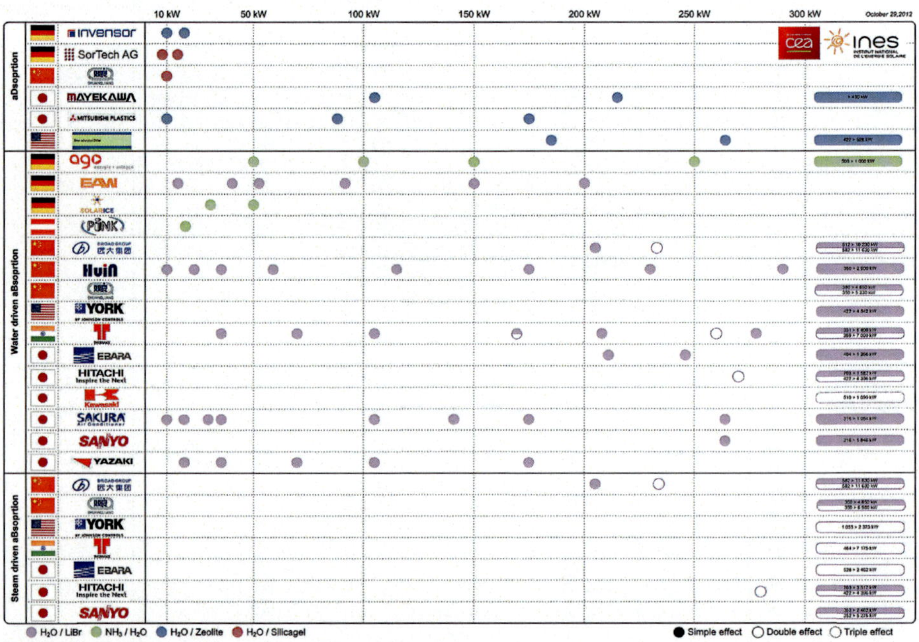

Figure 6.10 Thermally driven chiller market overview realized at end of 2012 by International Energy Agency solar heating and cooling Task 48.
Source: CEA INES.

a good overview of commercially available thermally driven chillers (closed systems), which was realized in the framework of the IEA SHC Task 48 "Quality Assurance & Support Measures for Solar Cooling Systems" at end of 2012 (note: some of the products have been changed/replaced in recent years).

6.3.1 Absorption chiller

Single-effect absorption chillers with the working pair water/lithium bromide or ammonia/water are generating cold using a closed, continuous cycle (Figure 6.11). In addition to this, there are also double-effect and triple-effect water/lithium bromide absorption chillers available on the market, which operate the same way but have higher efficiencies (two or three times higher than a normal single-effect chiller). The water/lithium bromide chillers are using water as refrigerant; hence the evaporator temperature is limited to temperatures around +5 °C. In contrast, the ammonia/water absorption chillers could generate evaporator temperatures down to −60 °C, which are useful for industrial cold processes. In absorption chillers, the refrigerant (water or ammonia) is absorbed by a liquid sorbent (lithium bromide or water). In the directly or indirectly solar/thermal-powered generator with high heating temperatures, the refrigerant is desorbed from the solution. This generates a high refrigerant vapor pressure, which is sufficient to condense the refrigerant in the condenser. After evaporation, the refrigerant vapor is absorbed in the solution that is cooled in the absorber. The solution is pumped to the generator by a solution pump in which it is regenerated and throttled back to the absorber. The heating temperatures for desorption are between 70 and 120 °C according to the technology. Basically, absorption chillers are used as central air-conditioning systems with decentralized fan coils or cooled ceilings.

Medium- and large-scale absorption chillers with cooling capacities above 35 kW have been available for several years. Today these chillers are mainly manufactured in China, India, Japan, and South Korea. In contrast, new small- and medium-scale absorption chillers have been developed in the past 10 years. These absorption chillers have cooling capacities from 10 to 500 kW_r.

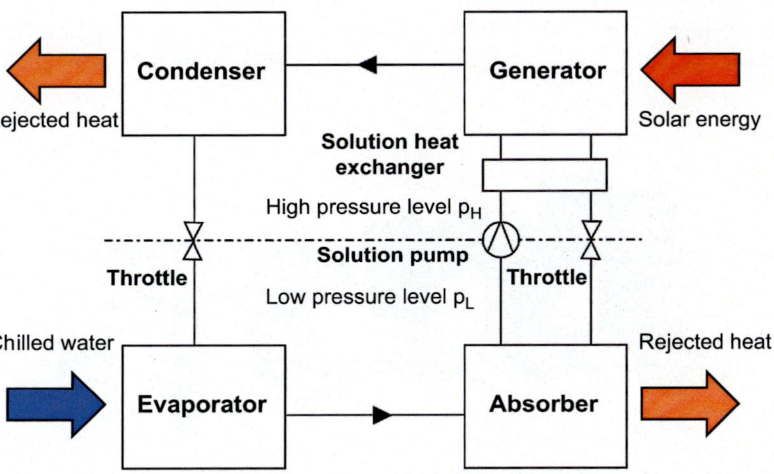

Figure 6.11 Functional principle of a single-effect absorption chiller.
Source: Dr Jakob Energy Research.

The company EAW from Germany has offered 15–200 kW_r water/lithium bromide (H_2O/LiBr) absorption chillers since 1998 (Figure 6.12). At a cold water temperature of 11 °C these absorption chillers produce a heat rejection temperature of 30 °C, hot water temperature at 90 °C, and a COP of 0.75. Low generator temperatures, in the range 70–90 °C, can be used, with correspondingly reduced cooling capacity. Other new German companies offering H_2O/LiBr chillers are Meibes System–Technik (since 2015, small-scale 5 kW_r) and Baelz (50 and 160 kW_r). Since 1977, the Japanese company Yazaki has offered H_2O/LiBr absorption chillers with a capacity range of 17.5–175 kW_r. The smallest absorption chiller with 17.5 kW_r cooling capacity provides cold water temperatures of 12.5/7 °C and a COP of 0.70 at driving temperatures of 88/83 °C and heat rejection temperatures of 31/35 °C. Another Japanese company is Sakura, which also offers H_2O/LiBr chillers in the capacity range of 10.5–316 kW_r. A further supplier of H_2O/LiBr absorber is the company Thermax from India. They offer chillers in capacity range of 17.5–352 kW_r. The COP of these chillers is about 0.70 at driving temperatures of 91/85 °C and heat rejection temperatures of 29/36.5 °C and cold water temperatures of 12.5/7 °C. The company Jiangsu Huineng, China, offers H_2O/LiBr absorption chillers from 11 to 350 kW_r. The heating temperature is 90/85 °C and heat rejection of 30 °C for cold water temperature of 15/10 °C. Jiangsu Huineng also produces an integrated absorption system, which is a combination of absorption chiller, wet cooling tower, and pumps in a single enclosure.

Small-scale ammonia/water (NH_3/H_2O) absorption chillers have been available since the end of 2006, with driving temperatures in the range of 65–95 °C. The Austrian company, Pink, offers one chiller with a cooling capacity of 19 kW_r at present (Figure 6.13). The chiller has a COP of 0.63 at driving temperatures of 85/75 °C, heat rejection temperatures of 24/31 °C and cold water temperatures of 12/6 °C. If water/glycol temperatures of 0/−3 °C are required, then cooling capacity is reduced to 10 kW_r, with a COP of 0.5. Other ammonia/water chillers have been available since 2008 from AGO in Germany, with 50–1000 kW_r (Figure 6.13), and from En-Save, also in Germany, with 30–100 kW_r capacity. Further ammonia/water absorption chillers are available from the German company Köhler Industries (40–250 kW_r).

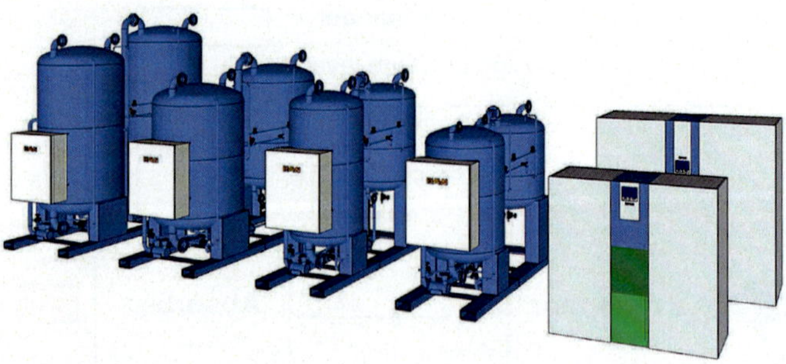

Figure 6.12 Latest water/LiBr absorption chillers for solar cooling systems. Source: EAW.

Figure 6.13 Small and medium-scale ammonia/water absorption chillers.
Sources: Pink (left), AGO (right).

6.3.2 *Adsorption chiller*

In adsorption chillers, the refrigerant water is adsorbed on a solid sorbent like silica gel or zeolite during disposal of latent heat on the surface. The latent heat decreases to zero with increasing addition of water molecules, and then only evaporative heat has to be dissipated. The desorption of the stored water and the pressure generation for the condensation is already caused by low heating temperatures of 50–70 °C, so that this technology is especially appropriate to the application of solar energy. The closed adsorption chillers are generating cold water of minimal 5–6 °C through the periodical cycle (Figure 6.14). These chillers can also be used as central air-conditioning systems with decentralized fan coils or cooled ceilings.

One problem of closed adsorption chillers is the poor heat transfer between the solid adsorbent material around a heat exchanger and the heat-transfer medium. Operation of adsorption chillers with very short cycle times (a few minutes) is beneficial in terms of the COP, but currently only possible using heat exchangers coated with adsorption material rather than packed adsorbent beds. The German company InvenSor has developed such direct-coated water/zeolite adsorption chillers; today, three chillers with a cooling capacity of 10, 18, and 30 kW_r are produced (Figure 6.15). The 18 kW_r chiller is intended for high ambient temperatures above 40 °C and a COP about 0.5, whereas the 10 and 30 kW_r chillers are developed for very low heating temperatures, starting at 50 °C for waste-heat applications and delivering a nominal COP of 0.6. Both chillers also include an integrated station producing and distributing cold water, which means that installation essentially comes down to the external pipework systems, as the necessary pumps, mixers, and valves are already included in the machine. The Japanese company Mayekawa also offers water/zeolite chillers with 105–430 kW_r cooling capacity. The cold-water temperatures are 16/9 °C, with heating and heat rejection temperatures of 75/67 °C and 29/34 °C, respectively. Since

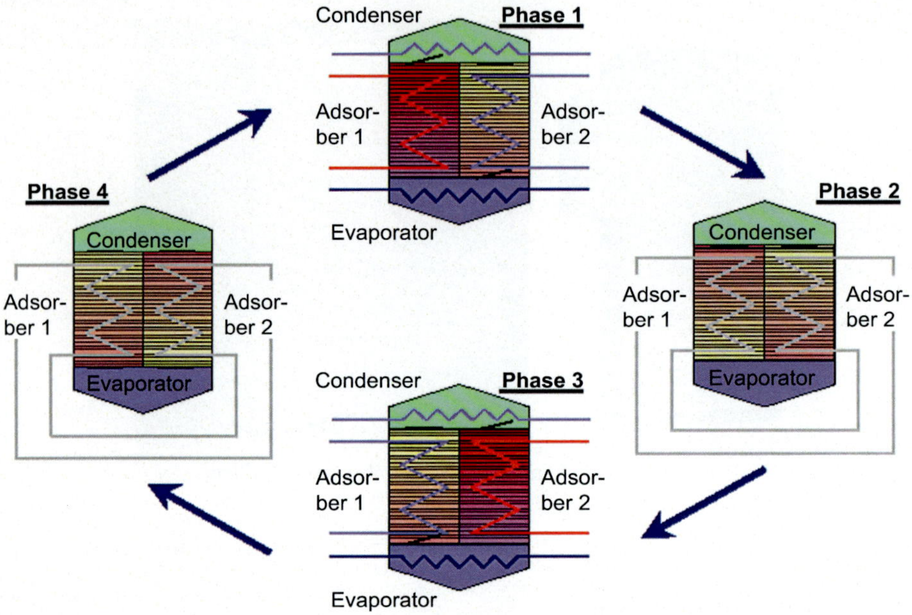

Figure 6.14 Functional principle of an adsorption chiller.
Source: Fraunhofer ISE.

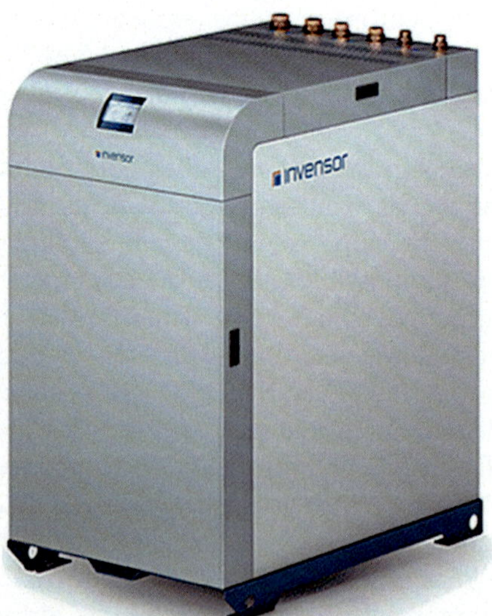

Figure 6.15 Small-scale water/zeolite adsorption chillers.
Source: InvenSor.

Figure 6.16 Small water/silica gel adsorption chillers for solar cooling kits.
Sources: SorTech (left), SolabCool (right).

2012, the Japanese company Mitsubishi Plastics has had two different 10 kW_r water/zeolite adsorption chillers on the market. One of the chillers comes with an integrated cooling tower in a single shell. Since 2015, the Germany company SorTech is also offering a 10 kW_r water/zeolite adsorption chiller.

SorTech offers also a water/silica gel adsorption chiller with 10 kW_r cooling capacity (Figure 6.16). They deliver cold water at temperatures of 18/15 °C, with a COP of 0.60 at driving temperatures of 72/65 °C and heat rejection temperatures of 27/33 °C (wet cooling tower). Using a dry cooler (33/38 °C) requires driving temperatures of 85/75 °C. The company HIJC from the USA also offers water/silica gel chillers with a higher cooling capacity of 220–350 kW_r. The company SolabCool does the latest development of a small-scale silica gel adsorption chiller in the Netherlands. They offer a 4.5 kW chiller with integrated heat rejection system (Figure 6.16).

6.3.3 Open sorption systems

Open sorption systems also called DEC systems supply conditioned air to a building, which is controlled to a specific temperature (lowest temperature is 16 °C) and humidity. The principle of open systems is to use the ambient air or a combination together with recirculated building air for air conditioning of a building (Figure 6.17) instead of chilled water. With such open systems building heat is removed by the air flow through the building, and additional fresh air is continuously supplied into the building. Therefore, air conditioning and building ventilation are provided at the same time. For the sorption part (solid-based sorption wheel or liquid-based salt solutions) of such an open system, solar heat is necessary for the regeneration to ensure a continuous operation.

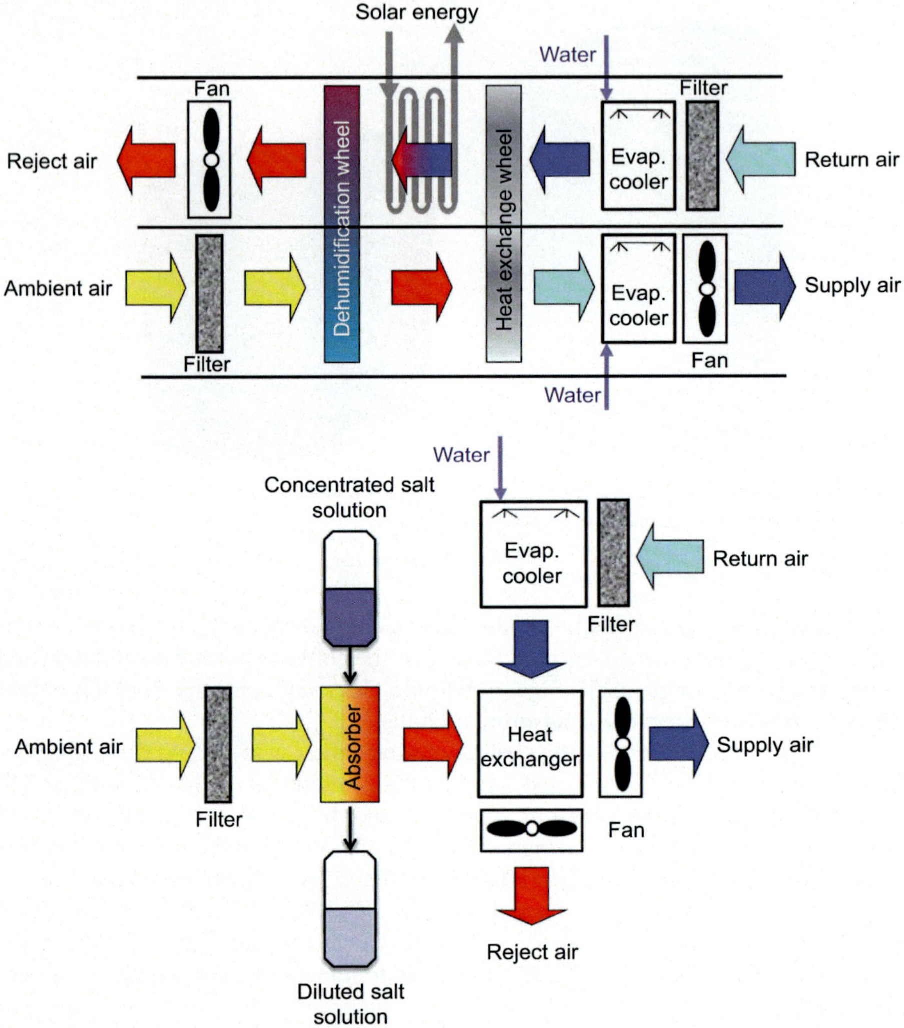

Figure 6.17 Functional principles of desiccant and evaporative cooling systems (top: solid sorption, bottom: liquid sorption).
Source: Solem Consulting.

The advantage of a solar heat-driven DEC system is that it fulfills all the essential requirements of air conditioning (i.e., control of fresh air temperature, humidity, and volume flow). According to the market survey conducted in the IEA-SHC Task 38 "Solar Air-Conditioning and Refrigeration" in 2009, solar heat-driven DEC systems have a low market share of the solar cooling market in comparison to absorption and adsorption chillers (Figure 6.9). Thereby, solid sorption materials are clearly dominating the applied DEC technology; only a few DEC systems have been identified that operate with liquid sorption material. In the IEA-SHC Task 48 a new successive

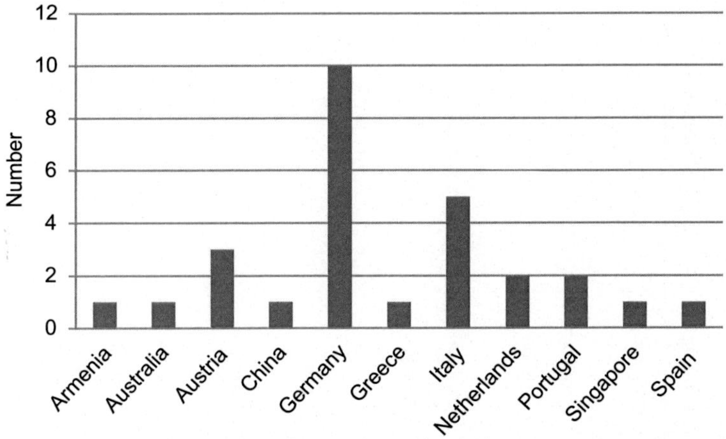

Figure 6.18 Identified desiccant and evaporative cooling systems per country realized in 2012 by International Energy Agency solar heating and cooling Task 48.
Source: AIT.

market screening was carried out in 2012. Twenty-eight operating DEC systems were identified. In terms of numbers, Germany, Italy, and Austria cover two-thirds of all the identified DEC installations (Figure 6.18). Based on the available information, only two DEC systems operate using liquid sorption material; all the other DEC systems use solid sorption material coating rotating matrices or fixed beds. The following companies offering solid DEC systems: e.g., Robatherm (Germany), Wolf Geisenfeld (Germany), and Munters (USA/Sweden). Liquid sorption system companies are AIL Research (USA), Imtech Drygenic (USA), L-DSC (Germany), and Menerga (Germany), which offer lithium chloride systems with 10–350 kW_r cooling capacity.

6.4 Solar cooling kit components

In recent years, some standardized systems have been developed as "solar cooling kits" to simplify and standardize the required technology and reduce the costs of solar cooling systems. In addition to solar energy, these kits are compatible with various thermal energy sources including biomass, CHP waste heat (combined heat and power), process heat, and district heating, or a combination of these sources. The aim of these product developments was to simplify and standardize the system technology and reduce the costs of solar cooling systems. The solar cooling kits offered by the companies generally contain flat-plate or evacuated-tube collectors, a hot water storage tank, solar station, pump set, sorption chiller, heat rejection unit, and a system controller. For example, there are established system suppliers for solar cooling kits in the power range up to 200 kW_r cooling capacity (Table 6.3). A detailed analysis of cost distribution for small to large solar cooling systems can be found in Kohlenbach and Jakob (2014).

Table 6.3 **System suppliers of solar cooling kits with up to 200 kW_r cooling capacity (status: 03/2015)**

System supplier	Country	Name of product	Nominal cooling capacity (kW_r)	Technology	Working pair
Gasokol	Austria	coolySun	15, 30, 54, 83, 150, 200	Absorption	Water/LiBr
			8, 15	Adsorption	Water/silica gel
Jiangsu Huineng	China	Solar central air conditioning	11, 23, 35, 58, 115, 175	Absorption	Water/LiBr
Kloben	Italy	SOLARTIK	17.5, 35, 70, 105	Absorption	Water/LiBr
Lucy solar	China	Solar air-conditioning system	11.5, 23, 35, 58, 115, 175	Absorption	Water/LiBr
EDF Optimal solutions	France	Package system	17.5, 35, 70, 105, 140, 210	Absorption	Water/LiBr
SolarNext	Germany	chillii® cooling kits	15, 17.5, 30, 35, 70, 105, 175	Absorption	Water/LiBr
			19, 50, 100	Absorption	NH_3/water
			10	Adsorption	Water/silica gel
			10, 18, 30	Adsorption	Water/zeolite
VICOT	China	Solar air-conditioning system	17.5, 35, 53, 70, 88, 105, 123, 141	Absorption	Water/LiBr

6.5 Research and development needs and future trends

So far, solar cooling is a niche product. Therefore, the challenge is market deployment and the difficulties in reducing costs while improving system quality and reliability despite the different technical and research and development developments that have been achieved or are currently ongoing. However, the main technical challenge for solar cooling today is at the system level. Many systems experience difficulties in achieving the planned energy savings and electrical performance (target $COP_{el} > 10$) because of shortcomings in proper system design and control that result in a high overall electricity consumption of the auxiliary components. In particular, this concerns the heat rejection unit and the fact that many systems are far too complex and, as a result, created massive control and maintenance efforts. To solve these technical barriers the following research and development topics must be tackled (Mugnier, 2011):

- Heat rejection: low electricity and water consumption with control capacity for part load. Hybrid systems (dry/wet) seem to be a promising solution.
- Highly efficient auxiliary components like pumps, careful hydraulic design, and advanced control systems. This is particularly important as solar cooling systems need more hydraulic loops than standard solutions.
- Integration of all components into a complete system: packaging efforts for small as well as large solar cooling systems.

The potential for further improvement of solar cooling energy and cost performance is most significant on the system level. Solar cooling is a complex technology and requires much more standardization for the coupling of the key components (solar thermal collectors and sorption chillers) and the development of robust, standardized solar cooling kits in the future. The industry and banking sector is starting to contribute here to provide turnkey solar cooling installations with guarantees on the performance as well as Energy Service Company (ESCO) financing models. Besides improvements on the system level, research and development on the component level will enhance overall system performance in terms of energy and cost. One focus is on quality procedure for designing, commissioning, monitoring, operating, and maintaining of solar cooling systems. This is an extremely important step (Henning, 2011) for overcoming most of the barriers that stand against the development of a mature market in the sector. Such quality procedures have to fix the steps toward uniform and consistent planning, providing information, recommendations, and minimum requirements for each of its steps. This will contribute to an increase in awareness and acceptance of the technology and positively impact the number of applications because it would strengthen the stakeholders' trust and thus enhance the market potential of solar cooling systems. Moreover, the introduction of concepts, such as guarantee of solar results or minimum energy consumption, will stimulate the interest of investors and create the boundary conditions of a correct financial risk assessment, crucial for contracting or energy service activities. The quality procedure will also work as a support tool for the policy implementation of the national renewable energy targets and as the base for subsidy mechanisms. As part of the IEA-SHC Task 48 "Quality Assurance & Support

Measures for Solar Cooling Systems" (task duration October 2011–March 2015) the above topics were tackled, and as a result many reports are now available for further information (http://task48.iea-shc.org).

Sources of further information

Further information regarding solar cooling can be found in the following books and websites:

- Kühn, A. (Ed.), 2013. Thermally Driven Heat Pumps for Heating and Cooling. Universitätsverlag der TU Berlin, ISBN: 978-3-7983-2596-8. (Link: opus4.kobv.de/opus4-tuberlin/files/3945/kuehn_annett_handbook.pdf).
- Henning, H-M., Mario Motto, M., Mugnier D. (Eds.), Solar Cooling Handbook—A Guide to Solar Assisted Cooling and Dehumidification Processes. Ambra Verlag, ISBN: 978-3-99043-438-3, 2013.
- Eicker, U. Low Energy Cooling for Sustainable Buildings. John Wiley & Sons Ltd, ISBN: 978-0-470-69744-3, 2009.
- Eicker, U. Solar Technologies for Buildings. John Wiley & Sons Ltd, ISBN: 0-471-48637-X, 2003.
- IEA-SHC Task 25 "Solar Assisted Air Conditioning of Buildings", http://archive.iea-shc.org/task25/
- IEA-SHC Task 38 "Solar Air-Conditioning and Refrigeration", http://task38.iea-shc.org
- IEA-SHC Task 48 "Quality Assurance & Support Measures for Solar Cooling Systems", http://task48.iea-shc.org
- IEA-SHC Task 53 "New Generation Solar Cooling & Heating Systems (PV or solar thermally driven systems)", http://task53.iea-shc.org

References

Grossman, G., 2002. Solar-powered systems for cooling, dehumidification and air-conditioning. Solar Energy 72 (1), 53–62.

Henning, H.M., August 2011. Solar Cooling Position Paper, IEA SHC Task 38 Solar Air-conditioning and Refrigeration. Germany, Freiburg.

IEA, 2012. Technology Roadmap Solar Heating and Cooling.

Jakob, U., November 6, 2012. Marktentwicklung und Rahmenbedingungen thermisch angetriebene Kältetechnik und solare Kühlung. In: Proceedings of the BMU Statusworkshop Solare Kühlung, Fraunhofer-forum Berlin, Germany.

Kohlenbach, P., Jakob, U., 2014. Solar Cooling: The Earthscan Expert Guide to Solar Cooling Systems. Routledge Chapman & Hall, ISBN: 978-0-415-63975-0.

Mouchot, A., 1987. Die Sonnenwärme und ihre industriellen Anwendungen. Olynthus Verlag.

Mugnier, D., May 5, 2011. Latest and future developments of solar cooling. In: Proceedings of the Second Annual Conference of the European Technology Platform on Renewable Heating and Cooling—RHC Conference, Budapest, Hungary.

Schubert, K., Dreyer, J., 1977. Kühlen mit Sonnenenergie. Dornier Post, No. 1, pp. 9–11.

Part Two

Enabling technologies, special applications and case studies

Thermal energy storage for renewable heating and cooling systems

7

Luisa F. Cabeza, Eduard Oró *
GREA Innovació Concurrent, Universitat de Lleida, Edifici CREA, Lleida, Spain

7.1 Introduction

7.1.1 Thermal energy storage

7.1.1.1 Principles and requirements

Thermal energy storage (TES) allows the storage of heat and cold for later use. TES is also known as heat or cold storage (Mehling and Cabeza, 2008; Cabeza, 2012). TES can aid in the efficient use and provision of thermal energy whenever there is a mismatch between energy generation and use. This mismatch can be in terms of time, temperature, power, or site (Dincer and Rosen, 2002). The potential advantages on the overall system performance are as follows (Mehling and Cabeza, 2008):

- Better economics—reducing investment and running costs
- Better efficiency—achieving a more efficient use of energy
- Less pollution of the environment and fewer CO_2 emissions
- Better system performance and reliability

The basic principle is the same in all TES applications. Energy is supplied to a storage system for removal and use at a later time (Dincer and Rosen, 2002). A complete process involves three steps (Figure 7.1): charging, storing, and discharging. In practical systems, some of the steps may occur simultaneously, and each step can happen more than once in each storage cycle (Gil et al., 2010).

Several factors have to be taken into consideration when deciding on the type and the design of any thermal storage system, and a key issue is its thermal capacity. However, selection of an appropriate system depends on many factors, such as cost−benefit considerations, technical criteria, and environmental criteria (Gil et al., 2010; Cabeza, 2012).

Recently, storage concepts have been classified as active or passive systems (Gil et al., 2010). An active storage system is mainly characterized by forced-convection heat transfer into the storage material. The storage medium itself circulates through

* Current address: Catalonia Institute for Energy Research, IREC, Jardins de les Dones de Negre, Sant Adriá de Besós, Barcelona, Spain.

Renewable Heating and Cooling. http://dx.doi.org/10.1016/B978-1-78242-213-6.00007-2

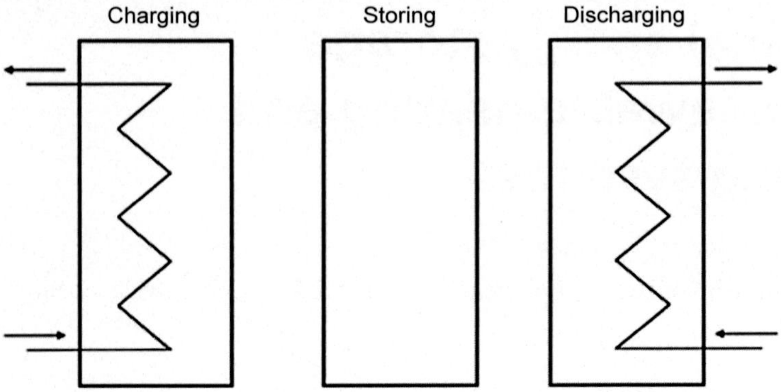

Figure 7.1 Steps involved in a complete TES system: charging, storing, and discharging (Gil et al., 2010).

a heat exchanger (the heat exchanger can also be a solar receiver or a steam generator). This system uses one or two tanks as storage media. Active systems are subdivided into direct and indirect systems. In a direct system, the heat transfer fluid (HTF) serves also as the storage medium, whereas in an indirect system, a second medium is used for storing the heat. Passive storage systems are generally dual-medium storage systems: the HTF passes through the storage only for charging and discharging a solid material.

The cost of a TES system mainly depends on the following items: the storage material itself, the heat exchanger for charging and discharging the system, and the cost of the space and/or enclosure for the TES.

From a technical point of view, the most important requirements are as follows:

- High energy density in the storage material (storage capacity)
- Good heat transfer between HTF and storage medium (efficiency)
- Mechanical and chemical stability of storage material (must support several charging–discharging cycles)
- Compatibility between HTF, heat exchanger, and/or storage medium (safety)
- Complete reversibility of a number of charging–discharging cycles (lifetime)
- Low thermal losses
- Easy control

And the most important design criteria from the point of view of technology are:

- Operation strategy
- Maximum load
- Nominal temperature and specific enthalpy drop in load
- Integration into the whole application system

7.1.1.2 Design of storages

Figure 7.2 shows the basic working scheme of heat storage: heat or cold supplied by a heat source is transferred to the heat storage, stored in the storage, and later transferred to a heat sink to cope with the demand (Mehling and Cabeza, 2008).

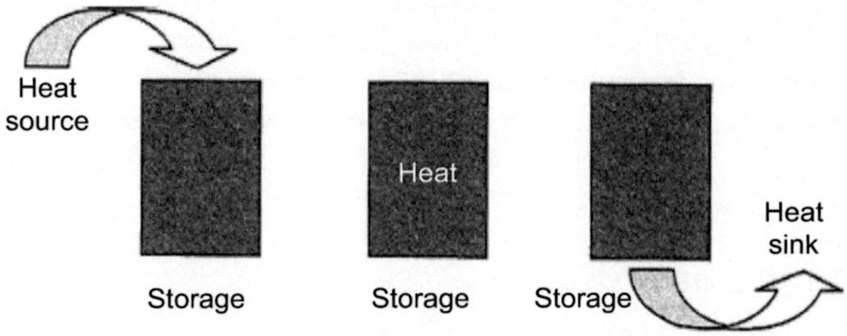

Figure 7.2 Basic working scheme of storage: heat or cold from a source is transferred to the storage, stored in the storage, and later transferred to a sink (Mehling and Cabeza, 2008).

Every application sets a number of boundary conditions, which must be carefully examined:

- From the temperature point of view, the supply temperature at the source has to be higher or equal to the temperature of the storage and the storage to the sink.
- From the power point of view—that is, the amount of heat transferred in a certain time must be that required in the charging and discharging.
- In some applications, the HTF and its movement by free or forced convection have to be considered.

There are three basic design options in storage systems (Mehling and Cabeza, 2008). The first one is when heat is exchanged by heat transfer on the surface of the storage. This becomes a typical heat transfer problem in which heat transfer resistance on the surface of the storage tank is the main parameter. Conduction and free or forced convection mechanisms are to be considered here.

Second, when a heat exchanger is used separating the HTF with the storage material, the surface of heat transfer increases significantly. This surface can be increased even further with the use of fins.

Finally, a third scheme is used when the heat storage medium is also the heat transfer medium. An example is when a water tank is discharged due to the demand of the shower, and cold water enters the tank replacing the hot one. In this case, heat transfer is basically by convection.

7.1.1.3 Integration of storage into systems

The main goal to integrate a heat or cold storage tank into a system is to supply heat or cold. However, the different supply and demand situations have a great influence on the integration concept (Mehling and Cabeza, 2008). The first case to consider is when there is no overlap in time between loading from the supply and unloading to the demand. In this case, the storage system can match different times of supply and demand; in many cases, the storage system can match different supply and demand power, and even supply and demand location, with transport of the storage medium. If there is a partial or total overlap in time, it is possible to smooth out fluctuations of the supply and/or the demand. Thus, the typical goals of storage integration are temperature regulation and power matching. The basic goals of the storage are to match supply and demand regarding the amount of heat and cold and the heating or cooling power at the right time. Although the amount of heat or cold is determined by the size of the storage and the heating or cooling power, which depend mainly on the design of the storage, the integration concept has a large influence with respect to time.

7.1.2 Methods for TES

7.1.2.1 Sensible heat storage

In sensible TES, energy is stored by changing the temperature of a storage medium such as water, air, oil, rock beds, bricks, concrete, or sand. The amount of energy introduced to the storage system is proportional to the temperature lift, the mass of the storage medium, and the heat capacity of the storage medium. Each medium or material has its own advantages and disadvantages, but usually its selection is based on the heat capacity and the available space for storage (Dincer and Rosen, 2002). The amount of heat stored in a material, Q, can be expressed as:

$$Q = m \cdot c_p \cdot \Delta T$$

in which m is the mass of storage material (kg), c_p is the specific heat of the storage material (J kg^{-1} °C^{-1}), ΔT is the temperature change (°C).

Sensible storage is the most common method of heat and cold storage. Some common materials used in TES systems are presented in Table 7.1 (Dincer and Rosen, 2002). The material must be inexpensive and should have good thermal capacity ($\rho\ c_p$, ρ is the density) to be useful in a storage application.

Besides the density and the specific heat of the storage material, other properties that are also important for sensible heat storage are operational temperatures, thermal conductivity and diffusivity, vapor pressure, compatibility among materials, stability, heat loss coefficient as a function of the surface area-to-volume ratio, and cost (Gil et al., 2010).

A sensible TES system consists of a storage medium, a container (commonly, a tank), and inlet–outlet devices. Tanks must retain the storage material and prevent losses of thermal energy. The existence of a thermal gradient across storage is desirable (Gil et al., 2010). Sensible heat storage can be made from solid or liquid media. Solid media are usually used in packed beds, requiring a fluid to exchange heat. When the fluid is a liquid, the heat capacity of the solid in the packed bed is not negligible, and the system is called a dual-storage system.

In heating and cooling of buildings, sensible TES is mostly used for domestic hot water, in combisystems (Hadorn, 2005), and for seasonal energy storage.

7.1.2.2 Latent heat storage

When a material stores heat while at phase transition, the heat is stored as latent heat. The solid–liquid phase change process by melting and solidification can store large amounts of heat and cold if a suitable material is selected. Upon melting, while heat is transferred to the storage material, the material still keeps its temperature constant

Table 7.1 Thermal capacity at 20 °C of some common materials used in sensible TES (Dincer and Rosen, 2002)

Material	Density ($kg\ m^{-3}$)	Specific heat ($J\ kg^{-1}\ K^{-1}$)	Volumetric thermal capacity ($\times 10^6$, $J\ m^{-3}\ K^{-1}$)
Clay	1458	879	1.28
Brick	1800	837	1.51
Sandstone	2200	712	1.57
Wood	700	2390	1.67
Concrete	2000	880	1.76
Glass	2710	837	2.27
Aluminum	2710	896	2.43
Iron	7900	452	3.57
Steel	7840	465	3.68
Gravelly earth	2050	1840	3.77
Magnetite	5177	752	3.89
Water	988	4182	4.17

at the melting temperature, also called phase-change temperature (Mehling and Cabeza, 2008). This is one of the main differences with sensible heat (Figure 7.3).

Usually the solid–liquid phase change is studied, but some solid–solid phase changes are of interest in some applications. The amount of heat stored can be calculated by:

$$Q = m \cdot \Delta h$$

in which m is the mass of storage material (kg) and Δh is the phase change enthalpy, also called melting enthalpy or heat of fusion ($J\ g^{-1}$).

Figure 7.4 shows the typical range of melting enthalpy and temperature of common material classes used as phase-change materials (PCMs) (Mehling and Cabeza, 2008). The best known and the mostly commonly used PCM is water, which has been used for cold storage since early times. For temperatures below 0 °C, water–salt solutions are the typically used materials. For temperatures between 0 and 130 °C, paraffins, salt hydrates, fatty acids, and sugar alcohols are used. For temperatures above 150 °C, salts and other inorganic materials are utilized. Many substances have been studied as potential PCMs, but only a few of them are commercialized (Zalba et al., 2003; Cabeza, 2005; Mehling and Cabeza, 2008; Cabeza et al., 2011). The selection of the material to be

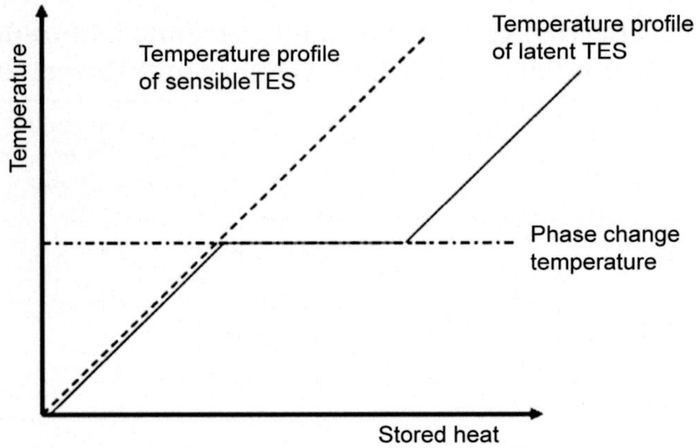

Figure 7.3 Heat storage as sensible and latent TES.

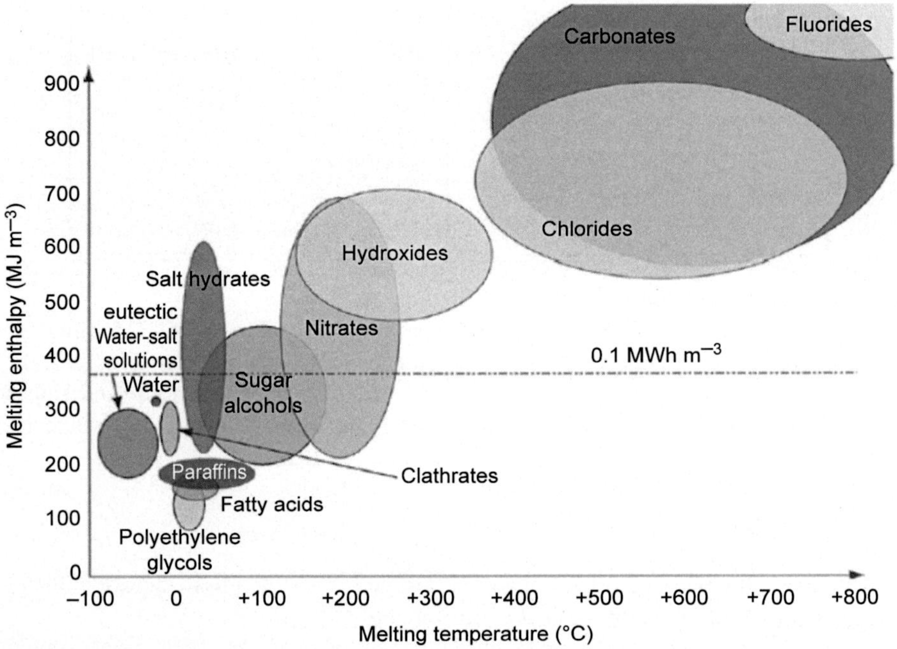

Figure 7.4 Classes of materials that can be used as PCMs and their typical range of melting temperature and melting enthalpy (Mehling and Cabeza, 2008).

used in latent heat storage is not easy. Availability and cost are usually the main drawbacks for the selection of a technically suitable material. Still today, problems such as phase separation, subcooling, corrosion, long-term stability, and low heat conductivity have not been totally solved and are under research.

Table 7.2 Comparison of organic and inorganic materials for heat storage (Zalba et al., 2003; Mehling and Cabeza, 2008)

	Organic	Inorganic
Advantages	No corrosives	Greater phase change enthalpy
	Low or no subcooling	
	Chemical stability	
Disadvantages	Lower phase change enthalpy	Subcooling
	Low thermal conductivity	Corrosion
	Flammability	Phase separation
		Phase segregation, lack of thermal stability

PCMs must have a large latent heat and high thermal conductivity, but the most important selection parameter is that the melting temperature of the materials lies in the practical range of operation. Other parameters are congruent melting, minimum subcooling, chemical stability, low cost, nontoxicity, and noncorrosivity. Materials that have been studied are paraffin waxes, fatty acids, and eutectics of organic and nonorganic compounds (Farid et al., 2004; Sharma et al., 2009). Tables 7.2 and 7.3 show two comparisons between organic and inorganic PCMs (Zalba et al., 2003; Mehling and Cabeza, 2008; Rathod and Banerjee, 2013).

According to Kenisarin and Mahkamov (2007), the following PCM properties to be used for latent heat storage were highlighted as desirable:

- High value of the heat of fusion and specific heat per unit volume and weight
- Melting point that matches the application
- Low vapor pressure (<1 bar) at the operational temperature
- Chemical stability and noncorrosiveness
- Not hazardous, highly inflammable, or poisonous
- Reproducible crystallization without degradation
- Small subcooling degree and high rate of crystal growth
- Small volume variation during solidification
- High thermal conductivity
- Availability and abundance

Paraffins are a mixture of pure alkanes that have quite a wide range of the phase-change temperature. These paraffins also have low thermal conductivity compared to inorganic materials, and therefore the choice of those which can be used for practical solar applications is very limited.

Commercial paraffin waxes are cheap with moderate thermal storage densities ($\sim$200 kJ kg^{-1} or 150 MJ m^{-3}) and a wide range of melting temperatures. They

Table 7.3 **Comparison of PCM types**

	Organic		Inorganic		
	Paraffins	**Fatty acids**	**Salt hydrates**	**Metals**	**Eutectics**
Formula	C_nH_{2n+2} ($n = 12–38$)	$CH_3(CH_2)\cdot COOH$	$AB\cdot nH_2O$	–	–
Melting point	−12–71 °C	7.8–187 °C	11–120 °C	30–96 °C	4–93 °C
Melting enthalpy	190–260 kJ kg^{-1}	130–250 kJ kg^{-1}	100–200 kJ kg^{-1}	25–90 kJ kg^{-1}	100–230 kJ kg^{-1}
Cost	Expensive	2 to 3 times more expensive than paraffins	Low cost	Costly	Costly

Adapted from Rathod and Banerjee (2013).

undergo negligible subcooling and are chemically inert and stable with no phase segregation. However, they have low thermal conductivity (~0.2 W m^{-1} K^{-1}), which limits their applications.

The main limitation of salt hydrates is their chemical instability when they are heated, as at elevated temperatures they degrade, losing some water content every heating cycle. Furthermore, some salts are chemically aggressive toward structural materials, and they have low heat conductivity. Finally, salt hydrates have a relatively high degree of subcooling.

Salt hydrates are attractive materials for use in TES due to their high volumetric storage density (~350 MJ m^{-3}), relatively high thermal conductivity compared to organic materials (~0.5 W m^{-1} K^{-1}), and moderate costs compared to paraffin waxes, with few exceptions.

According to Cabeza et al. (2011), the PCM to be used in the design of a thermal storage system should have desirable thermophysical, kinetic, and chemical properties and desired economics as listed below:

1. Thermophysical properties
   - **a.** Melting temperature in the desired operating temperature range: to assure storage and extraction of heat in an application with a fixed temperature range
   - **b.** High latent heat of fusion per unit volume: to achieve high storage density compared to sensible storage
   - **c.** High specific heat to provide additional significant sensible heat storage
   - **d.** High thermal conductivity of both solid and liquid phases to assist the charging and discharging energy of the storage system
   - **e.** Small volume change on phase transformation and small vapor pressure at operating temperature to reduce the containment problem
   - **f.** Congruent melting of the PCM for a constant storage capacity of the material with each freezing/melting cycle
   - **g.** Reproducible phase change: to use the storage material many times (also called cycling stability)
2. Kinetic properties—nucleation and crystal growth
   - **a.** High nucleation rate to avoid supercooling of the liquid phase and to assure that melting and solidification proceed at the same temperature
   - **b.** High rate of crystal growth, so that the system can meet demand of heat recovery from the storage system
3. Chemical properties
   - **a.** Complete reversible freeze/melt cycle
   - **b.** No degradation after a large number of freeze/melt cycles
   - **c.** No corrosiveness to the construction materials
   - **d.** Nontoxic, nonflammable, and nonexplosive material for safety: for environmental and safety reasons
4. Economics
   - **a.** Abundant
   - **b.** Available
   - **c.** Cost-effective: to be competitive with other options for heat and cold storage

For their use, PCMs must be encapsulated, either encapsulating the material or encapsulating the building composite, as otherwise the liquid phase would be able to flow away from the location where it is applied.

Latent heat storage is used in buildings both in passive and in active systems to reduce the energy demand of the building, to better use renewable energies (solar energy), or for free cooling.

7.1.2.3 Thermochemical heat storage

Any chemical reaction with high heat of reaction can be used for TES if the products of the reaction can be stored and if the heat stored during the reaction can be released when the reverse reaction takes place (Mehling and Cabeza, 2008). A comparison of the energy storage densities achieved with different methods of storage is shown in Table 7.4.

Higher energy storage density and reversibility are required on the materials for thermal energy conversion and storage (Kato, 2007). Energy density of chemical changes is relatively higher than that of physical changes. A merit of chemical energy conversion is the possession of efficient energy storage performance. The performance is especially advantageous for TES. Chemical storage can store energy as reactants with small loss. It is important to find the appropriate reversible chemical reaction for the temperature range of subjected energy source.

TES can be realized by utilizing reversible chemical reactions (Hauer, 2007). Here, the process of adsorption on solid materials or absorption on liquids is explained. Adsorption means binding of a gaseous or liquid phase of a component on the inner surface of a porous material. During the desorption step, heat is put into the sample. The adsorbed component is removed from the inner surface. As soon as the reverse reaction (adsorption) is started, the heat will be released. The adsorption step represents the discharging process. There are two types of sorption systems, closed and open storage systems. In a "closed sorption system," the heat is transferred to and from the adsorbent by a heat exchanger, usually called condenser/evaporator. The heat has to be transported to the absorber at the same time as it is extracted from the condenser to keep the HTF, usually water, flowing from the adsorber to the condenser.

The mechanism of a sorption thermal storage process can be represented by:

$$\mathrm{A}\cdot(m+n)\mathrm{B} + \mathrm{Heat} \leftrightarrow \mathrm{A}\cdot m\mathrm{B} + n\mathrm{B}$$

in which A is the sorbent and B is the sorbate. A/B is called a sorption working pair.

Although sorption thermal storage systems offer some benefits, there are still critical drawbacks, such as great complexity in the system configuration (for closed systems), expensive investment, poor heat and mass transfer ability (for chemical reaction), and low heat storage density in actual systems (Yu et al., 2013).

According to the system configuration, sorption storage systems can be divided into open and closed systems (Hauer, 2007; Yu et al., 2013). Closed sorption systems are not in contact with the atmospheric environment and have been studied for refrigeration, heat pump systems, and energy storage applications (Figure 7.5). Closed systems are adequate for small-scale applications, in which compact and highly efficient devices are needed. Open systems allow the release of the sorbate to the environment

Table 7.4 **Comparison of storage densities of different TES methods**

Type of storage technology	Material	Energy stored ($MJ\ m^{-3}$)	Energy stored ($kJ\ kg^{-1}$)	Comments
Sensible heat	Granite	50	17	$\Delta T = 20\ °C$
	Water	84	84	$\Delta T = 20\ °C$
Latent heat	Water	306	330	$T_{melting} = 0\ °C$
	Paraffins	180	200	$T_{melting} = 5–130\ °C$
	Salt hydrates	300	200	$T_{melting} = 5–130\ °C$
	Salt	600–1500	300–700	$T_{melting} = 300–800\ °C$
Chemical reactions	H_2 gas (oxidation)	11	120,000	300 K, 1 bar
	H_2 gas (oxidation)	2160	120,000	300 K, 200 bar
	H_2 liquid (oxidation)	8400	120,000	20 K, 1 bar
	Fossil gas	32		300 K, 1 bar
	Gasoline	33,000	43,000	
Electrical storage	Zn/Mn oxide battery		180	
	Pb battery		70–80	

Adapted from Mehling and Gabeza (2008).

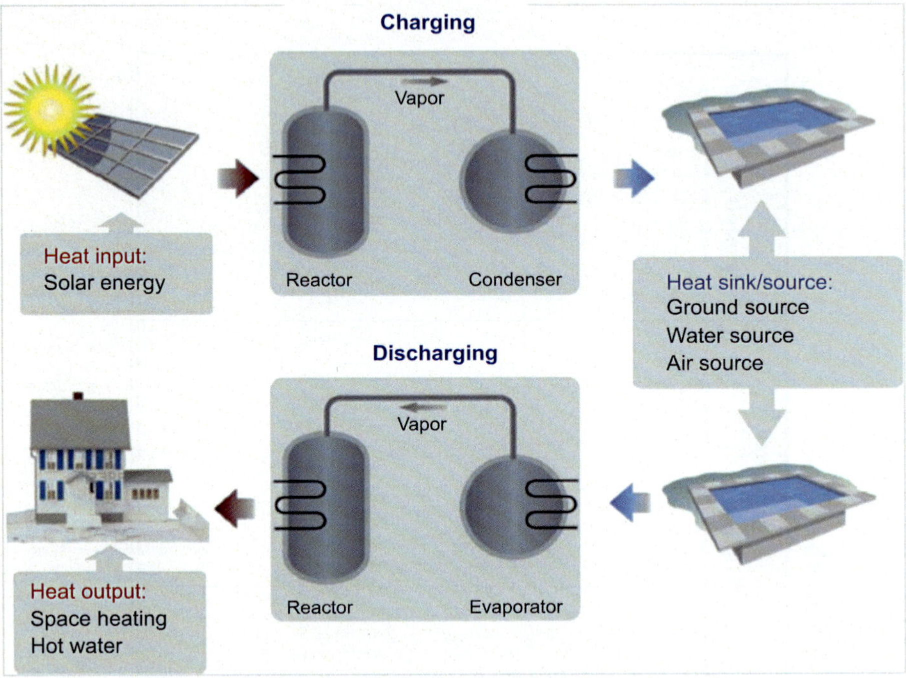

Figure 7.5 Closed sorption thermal energy storage system (Yu et al., 2013).

(Figure 7.6); the sorbate usually is water. These systems have lower investment costs and better heat and mass transfer than closed systems. Thermochemical storage is mostly studied for seasonal TES.

According to Yu et al. (2013), storage materials for thermochemical storage should achieve the following requirements:

- High energy storage density (kWh m^{-3})
- Low charging temperature
- High uptake of sorbate (g sorbate · g $sorbent^{-1}$)

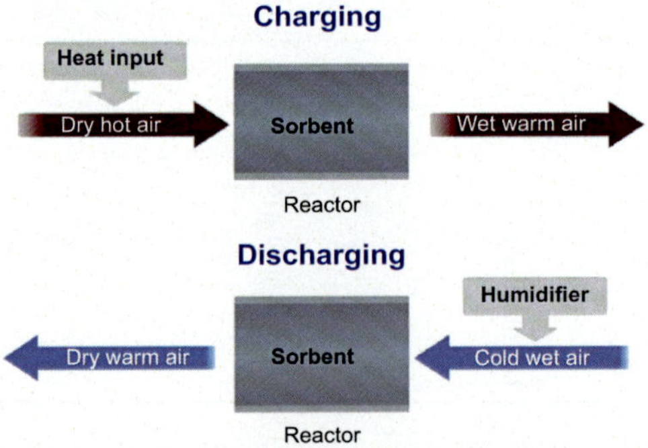

Figure 7.6 Operation principle of an open sorption thermal energy storage system (Yu et al., 2013).

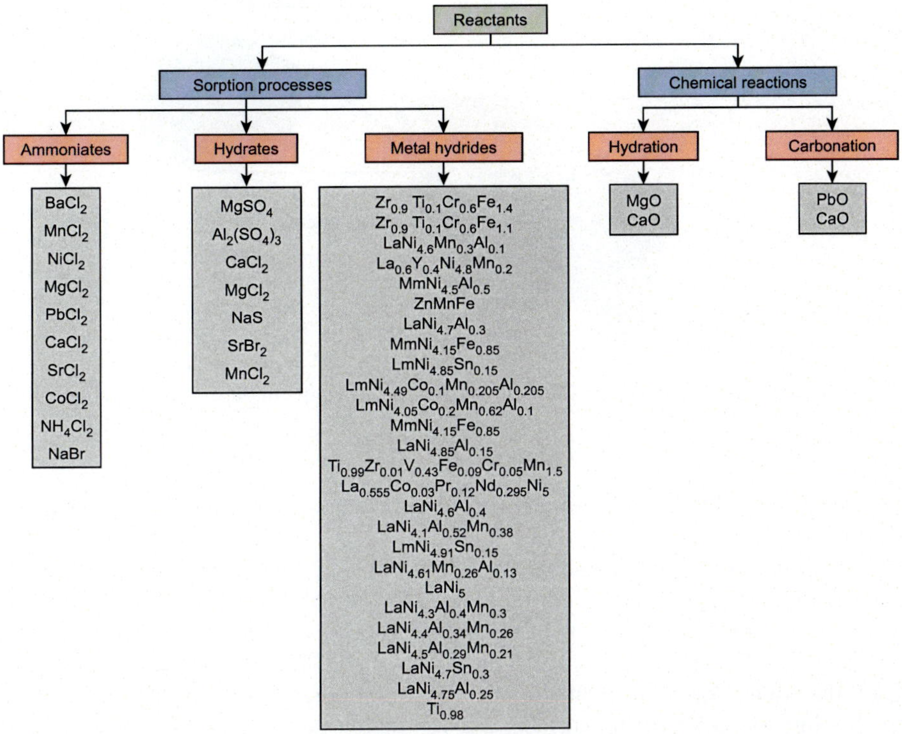

Figure 7.7 Working pairs tested for thermochemical storage (Cot-Gores et al., 2012).

- Appropriate heat and mass transfer properties to ensure designed output power
- Easy to handle, nonpoisonous
- Low cost, low price per kWh heat energy stored
- Thermal stability, no deterioration

The chemical sorption processes found in the literature are those between metal salts with water, ammonia, methanol or methyl-ammonia, and metal alloys with hydrogen (Figure 7.7). Sorption enthalpy typically ranges between 20 and 70 kJ mol^{-1}. A good description of the specific reactants can be found in Cot-Gores et al. (2012).

7.2 Description of TES technologies used today and their applications in the context of renewable heating and cooling

7.2.1 *Passive systems in building skins*

Passive use of PCMs in buildings has the objective to decrease the operation energy of the building by decreasing the energy demand of space heating and cooling, basically

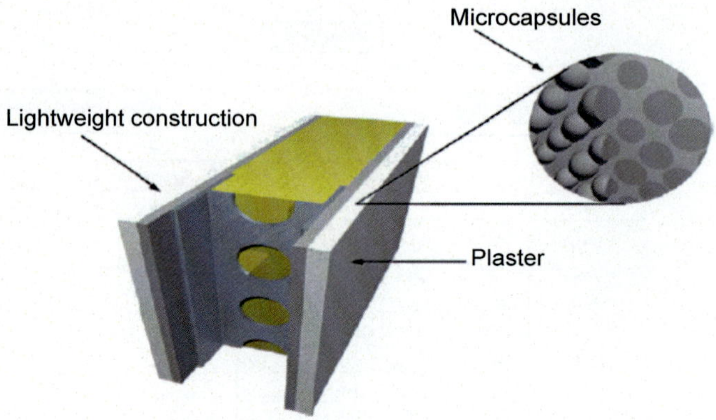

Figure 7.8 Addition of microencapsulated PCMs in a lightweight building (Schossig et al., 2005).

smoothing the indoor temperature by increasing the energy inertia of the building envelope.

One of the oldest options studied and published were PCM wallboards to improve the thermal comfort of lightweight buildings (Sharma et al., 2009), because they are very suitable for the incorporation of PCM (Soares et al., 2013). The efficiency of these elements depends on several factors such as how the PCM is incorporated in the wallboard, the orientation of the wall, the climatic conditions, the direct solar gains, the internal gains, the color of the surface, the ventilation rate, the PCM chosen, and its phase-change temperature, the temperature range over which phase change occurs, and the latent heat capacity per unit area of the wall.

Schossig et al. (2005) showed that PCM can be impregnated into building materials but that when added microencapsulated leakage problems disappear (Figure 7.8). Barreneche et al. (2013) compared in an interlaboratory test the three different methods to incorporate PCM in building materials: microencapsulated, suspension, and impregnation. Although suspension allows the maximum amount of PCM in the building material, only microencapsulation ensures avoiding leakage.

An example of the use of PCM in wallboards was the development of the Dupont product Energain, studied and characterized in several papers (Kuznik et al., 2008a,b; Kuznik and Virgone, 2009; Kuznik et al., 2011). Kuznik et al. (2008a) performed an optimization process using interior/exterior temperature evolutions within a period of 24 h to optimize the thickness of a PCM wallboard to enhance the thermal behavior of a lightweight internal partition wall. The PCM wallboard was composed of 60 wt% of microencapsulated paraffin, which has a melting temperature of about 22 °C (Figure 7.9).

PCM wallboard is considered to be an effective and less costly replacement of standard thermal mass to store solar heat in buildings, in which the PCM is imbedded into a gypsum board, plaster, or other building structures. The thermal characteristics of PCM wallboard are very close to those of PCMs alone, and when a PCM wallboard

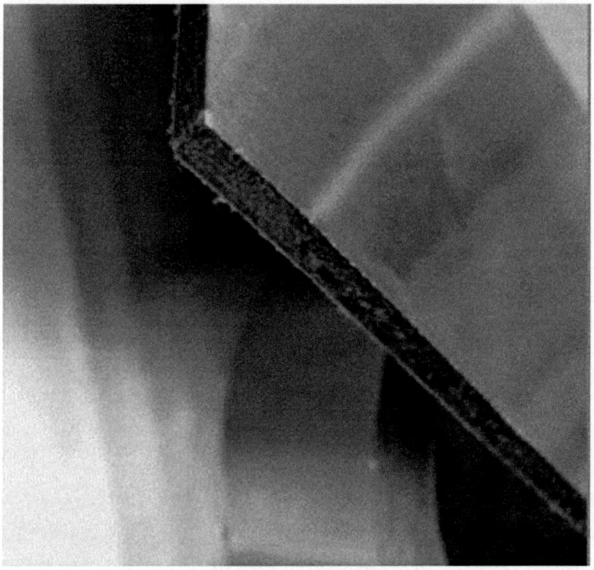

Figure 7.9 Dupont PCM composite wallboard (Kuznik et al., 2008a).

is cut, a greater concentration of PCM lies in the outer third of the wallboard thickness near each face due to the diffusion process (Athienitis et al., 1997).

Athienitis et al. (1997) used a gypsum board impregnated with a PCM in a direct-gain outdoor test room to investigate the thermal performance of PCM gypsum board used in a passive solar building. The results showed that the room temperature can be reduced by a maximum 4 °C during the daytime. Neeper (2000) reported on the impregnation of fatty acid and paraffin waxes into the gypsum wallboard and examination of the thermal dynamics under the diurnal variation of room temperature (the radiation absorbed was not considered) with the PCM on the interior partition and on the exterior partition, respectively. The investigation indicated that when the PCM melting temperature was close to the average room temperature, the maximum diurnal energy storage occurred and diurnal energy storage decreased if the phase change transition occurred over a range of temperature.

To evaluate the capacity of PCM to stabilize the internal environment when there were external temperature changes and solar radiations, Kuznik et al. (2008b) designed an experimental test room MINIBAT using a battery of 12 spotlights to simulate an artificial sunning and they got the results that the PCM wallboard can reduce the air-temperature fluctuations in the room and enhance the natural convection mixing of the air, avoiding uncomfortable thermal stratifications. Kuznik and Virgone (2009) also tested two identical test cells under two kinds of external temperature evolutions, heating and cooling steps with various slopes, and sinusoidal temperature evolution with 24 h period. They found there was time lag between indoor and outdoor temperature evolutions and the external temperature amplitude in the cell was reduced.

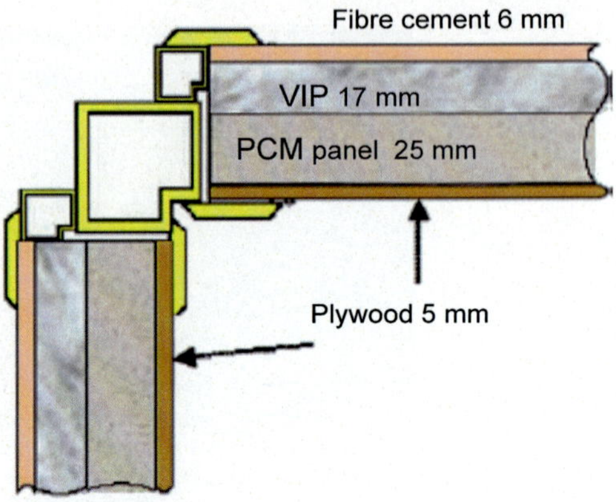

Figure 7.10 Cross-structure of PCM wallboard with vacuum isolation panel (VIP) (Ahmad et al., 2006).

Lv et al. (2006) built an ordinary room as well as a room using PCM gypsum wallboard in the northeast of China, and they found that the PCM wallboards can attenuate indoor air fluctuation, reduce the heat transfer to outdoor air, and have the function to keep warm. Recently, Kuznik et al. (2011) used Dupont de Nemours PCM wallboards for the renovation of a tertiary building and found they were really efficient if the outside temperature was varying in melting temperature by monitoring the building for a whole year. Some researchers reported that using a vacuum isolation panel (VIP) in a wallboard can reduce the thermal loss and improve efficiency for lightweight buildings.

Two test cells were designed by Ahmad et al. (2006), and each cell consisted of one glazed face and five opaque faces insulated with VIPs. One of the cells was equipped with five PCM panels. The cross-structure with PCM wallboard and VIP is shown in Figure 7.10. The amplitude of temperature variation inside the cell with PCM panels was decreased by 20 °C. So in the winter, it helped to efficiently prevent negative indoor temperature. The PCM panels still showed a good thermal storage capability even after more than 480 thermal cycles.

Mandilatas et al. (2013) present one of the first attempts to investigate the thermal performance of a purposely built full-scale lightweight demonstration house constructed in Greece that includes PCM gypsum boards in all external walls as well as in internal partitions of the building (Figures 7.11 and 7.12). Experimental results show that the indoor air temperatures in all thermal zones examined (LVR, MBDR, BDR) do not significantly vary during a 24 h day—night cycle and this can be attributed to the house's enhanced thermal mass associated with the insulation, as well as with the absence of typical occupant behavior.

The use of PCM in building envelopes has been demonstrated in a long-term experimental study at the University of Lleida (Spain). Here, different forms of PCM have been tested in a pilot plant (Figure 7.13) in several identically shaped cubicles with internal dimensions of 2.4 × 2.4 × 2.4 m. The cubicles were built using different typical constructive solutions so concrete cubicles (Cabeza et al., 2007), brick cubicles, and

Figure 7.11 External view of a demonstration house including PCM wallboards (Mandilaras et al., 2013).

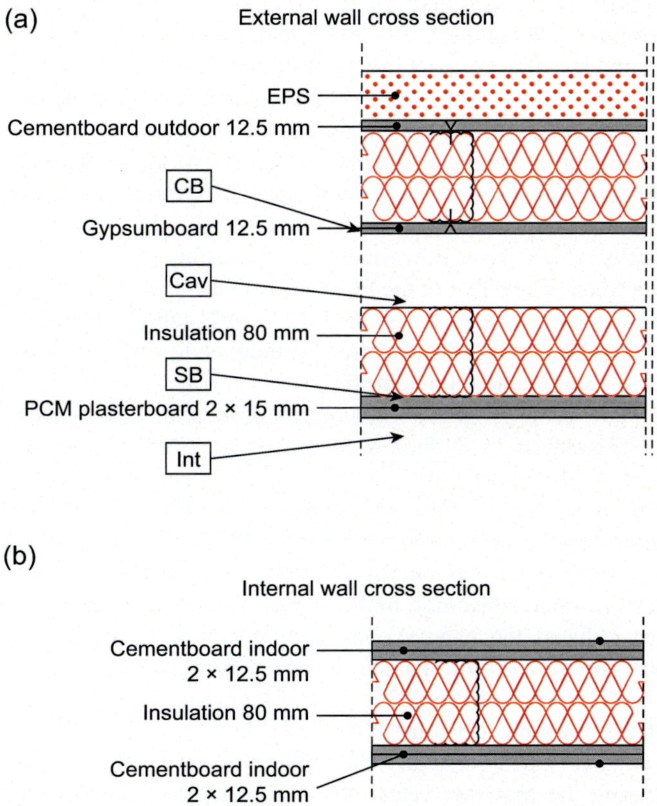

Figure 7.12 Cross section of (a) the external wall ("CB," "Cav," "SB," and "Int" correspond to temperature sensors placed in the LVR east wall); (b) the partition wall with cement boards (Mandilaras et al., 2013).

Figure 7.13 Experimental setup located in Puigverd de Lleida, Spain (Castell et al., 2010).

alveolar brick cubicles (Castell et al., 2010) can be found. In this setup, free-floating temperature and controlled temperature experiments are carried out.

Castellón et al. (2009) show different results of this experimental setup. As an example, results from the brick cubicles are shown here. Figure 7.14 presents the results for free-floating experiments in the brick cubicles, comparing the Reference, PU, and RT27 + PU cubicles during a week in August 2008, when the PCM was working within the phase-change range. As expected, the Reference cubicle always presented higher temperature oscillations and it was also more sensitive to the ambient temperature changes than the PU and RT27 + PU cubicles. When comparing the PU and the RT27 + PU cubicles, the temperature control achieved by the use of PCM was observed. The temperature of the RT27 + PU cubicle remained closer to the phase-change range. The RT27 + PU cubicle remained cooler (about 0.4 °C) during most of the week, when the weather was warmer. At the end of the week (when the weather is cooler) the tendency was reversed, with the PU cubicle being cooler than the RT27 + PU cubicle. This behavior can be explained by the higher thermal mass of the PCM cubicle, which slows down the general cooling tendency that occurs in the last days of the week. The effect of the PCM is also visible in the reduction in the daily oscillations of the inside temperature in the RT27 + PU cubicle. It is also observed that the effect of the PCM is only partially used, as there is no single 24 h period in which full melting and solidification are achieved.

Figure 7.15 and Table 7.5 present the results of the controlled temperature experiments using a set point of 24 °C for a week in August, 2008. The accumulated energy consumption of the Reference cubicle is much higher than all the other cubicles, with about twice the consumption of the other cubicles. The RT27 + PU cubicle is the one with the lowest energy consumption, whereas the SP25 + Alveolar cubicle is the second one, consuming even less energy than the PU cubicle. Finally, the Alveolar cubicle is the one that consumes more energy after the Reference cubicle. Both PCM cubicles reduced the energy consumption compared with the same cubicle without PCM. The RT27 + PU cubicle achieved a reduction of 14.75% compared with the PU cubicle, whereas the SP25 + Alveolar cubicle reached 17.12% of energy savings compared with the Alveolar cubicle (Table 7.5).

Investigations on PCM floors and PCM ceilings for passive solar heating have been carried out during the past few years. Xu et al. (2005) used shape-stabilized PCM floors in passive solar buildings and developed a model to analyze how various factors influence the thermal performance, such as thickness of PCM layer, melting

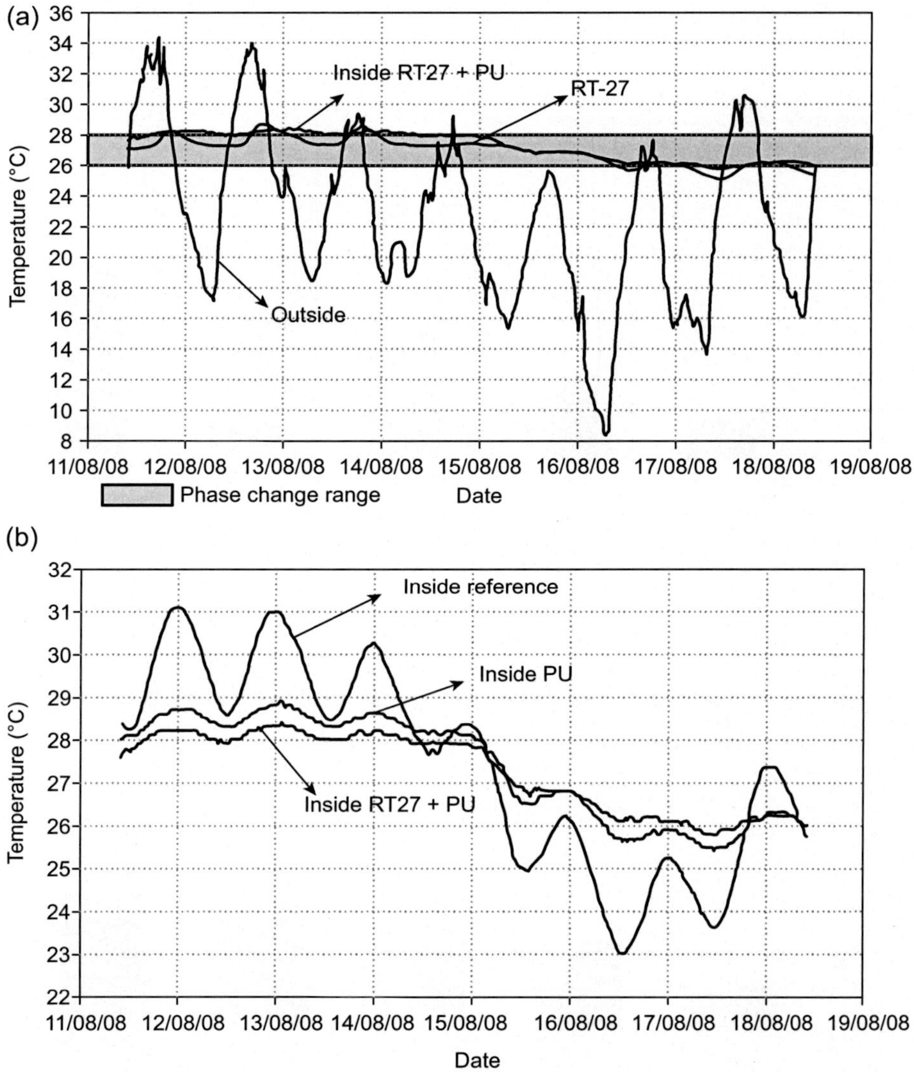

Figure 7.14 Brick cubicles free-floating experimentation: (a) Weather conditions and PCM operating temperature; (b) indoor ambient temperature for Reference, PU, and RT27 + PU cubicles (Castellón et al., 2009).

temperature, heat of fusion, and thermal conductivity of PCM. They indicated that the heat of fusion and thermal conductivity of PCM should be larger than 120 kJ kg^{-1} and 0.5 W m^{-1} K^{-1} and that the thickness of shape-stabilized PCM plate should not be larger than 20 mm. Pasupathy and Velraj (2008) studied the effect of the building with a PCM panel on the roof from the aspect of the location and thickness. They recommended a double-layer PCM to be incorporated in the roof to narrow indoor air temperature variation and to better suit it for all seasons.

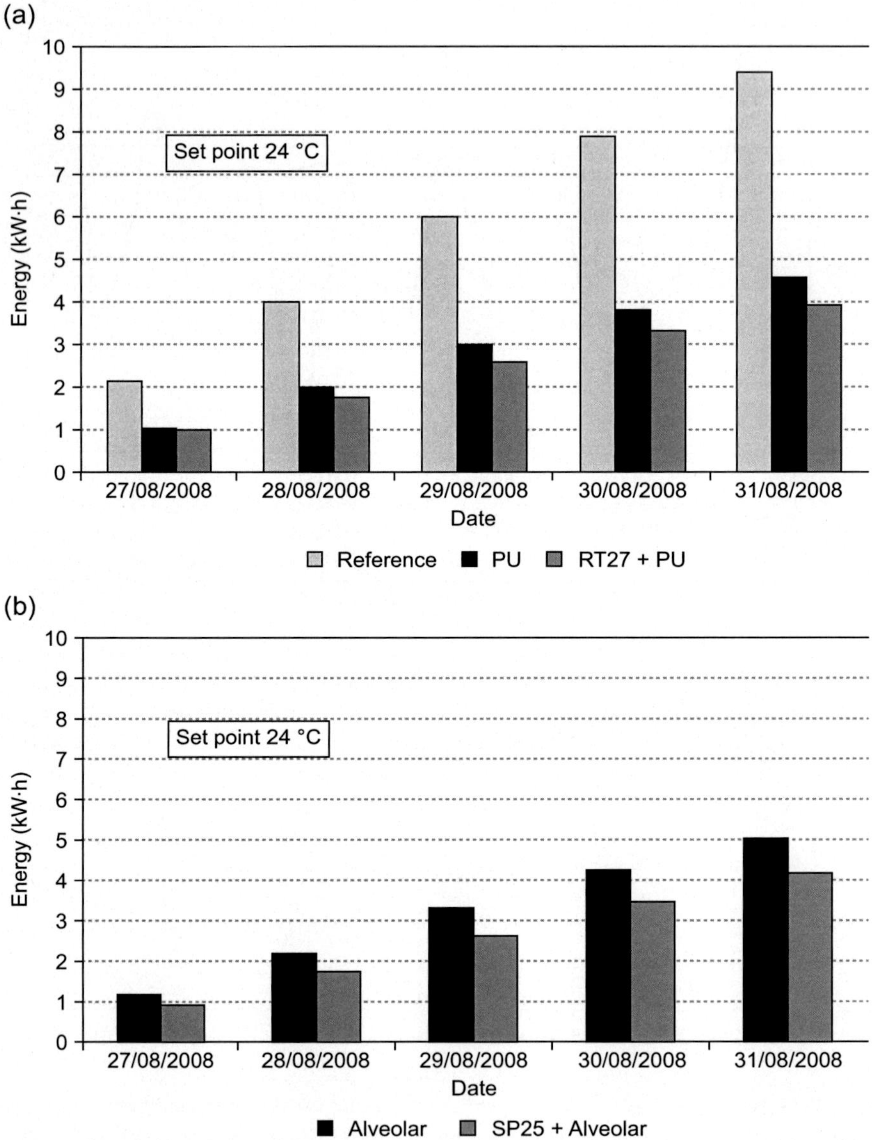

Figure 7.15 Energy consumption of the heat pumps in cooling mode for 5 days in August 2008: (a) Reference, PU, and RT27 + PU; and (b) Alveolar and SP25 + Alveolar cubicles (Castellón et al., 2009).

Many other examples of passive solar systems using PCM can be found in the literature and summarized in the different reviews published (Zalba et al., 2003; Farid et al., 2004; Kenisarim and Mahkamov, 2007; Cabeza et al., 2011; Sharma et al., 2009; Rathod and Banerjee, 2013; Soares et al., 2013).

Table 7.5 Accumulated energy consumption and energy savings for the different cubicles (Castell et al., 2010)

	Energy consumption (Wh)[a]	Energy savings (Wh)[b]	Energy savings (%)[b]	Improvement (%)[c]
Reference	9376	0	0	–
PU	4583	4793	51.1	0
RT27 + PU	3907	5469	58.3	14.8
Alveolar	5053	4323	46.1	0
SP25 + Alveolar	4188	5188	55.3	17.1

[a]Set point of 24 °C for 5 days.
[b]Referred to the reference cubicle.
[c]Referred to the cubicle with analog constructive solution and w/o PCM.

7.2.2 Active systems

Active systems using TES in buildings have the aim to decrease the operational energy of the building by decreasing the use of fossil fuels in heating, cooling, and domestic hot water production.

7.2.2.1 Water heating systems

Water storage tanks are made of a wide variety of materials, such as steel, aluminum, reinforced concrete, and steel (Tatsikjoudoung et al., 2013). Moreover, they are usually insulated with glass wool, mineral wool, or polyurethane. Water storage tank energy efficiency and performance are improved when stratification is increased.

In solar water heating systems, the use of PCM can be an advantage because the volume of the necessary water storage tank can be decreased (Cabeza et al., 2006). The PCM module geometry adopted in this study was to use several cylinders at the top of the water tank. A granular PCM–graphite compound of about 90 vol% of sodium acetate trihydrate and 10 vol% graphite was chosen as the PCM for the experiments presented here. The experiments presented in this paper showed that the inclusion of a PCM module in water tanks for domestic hot water supply is a very promising technology. It would allow having hot water for longer periods of time even without exterior energy supply or to use smaller tanks for the same purpose.

This concept of adding PCMs in solar water systems is quite controversial in the bibliography, because although results such as those from Cabeza et al. (2006) show the benefits, Talmatsky and Kribus (2008) carried out a numerical study in which annual simulations were done for different sites, load profiles, different PCM volume fractions, and different kinds of PCMs, showing that, contrary to the expectation, the use of PCMs in the storage tank does not yield a significant benefit in energy provided to the end user. Later, Kousksou et al. (2011) confirmed these findings regarding the use of PCMs but claimed that it seems highly desirable to incorporate a mathematical optimization at the early stages of the design process to achieve more realistic results.

The other big application of water storage systems is the combisystem, in which DHW and heating are produced. Streicher and Bales (2005) stated that one of the key elements of a solar heating system is the hot water store, which has to fulfill several tasks:

- Deliver sufficient energy to the heat sink
- Decouple mass flows of heat sources and heat sinks
- Store heat from unsteady heat sources (solar) from times when excess heat is available to times when too little or no heat is available (either short-term storage from day to night or over one to a few days, or seasonal storage)
- Extend the running times for auxiliary heating devices to increase their efficiency and lower startup/shutdown emissions
- Allow a reduction in heating capacity of auxiliary heating devices
- Store the heat at the appropriate temperature levels without mixing (stratification) to avoid energy losses

The design of the water store in a combisystem greatly affects the overall system performance. The principle of a water store connected to a solar system and with auxiliary heat input is shown in Figure 7.16. In these systems, the stratification of the hot water is of great importance, and charging and discharging must be done carefully considering this fact. This is why water tanks use stratifying units (Figure 7.17).

7.2.2.2 PCM in HVAC systems

The use of night cooling ventilation in addition to PCM is a very powerful strategy for reducing the cooling demand of buildings. Nevertheless, there are inherent drawbacks in the way things have been done so far:

- The limited area of contact between PCM and the air
- The very low convective heat transfer coefficient, which prevents the use of significant amount of PCM
- The very low utilization factor of the cool stored due to the large phase shift between the time when cool is stored and when it is required by the building

A very powerful well-known strategy for reducing the cooling demand of buildings is the use of low outdoor air temperatures during the night to cool the structures of the building. In many cases, the nighttime outdoor air climatic conditions of many Mediterranean locations allow a significant compensation of the daytime solar and internal heat gains (Allard et al., 1998).

The performance of a night cooling application depends on (Alvarez et al., 1997):

- The climate, which provides the availability of the heat sink in terms of its thermal level and its variability throughout the year and on a daily basis.
- The cooling needs of the building (absolute values and load profiles). The combination of 1 and 2 gives the extent to which the requirements can be covered by the cooling technique.

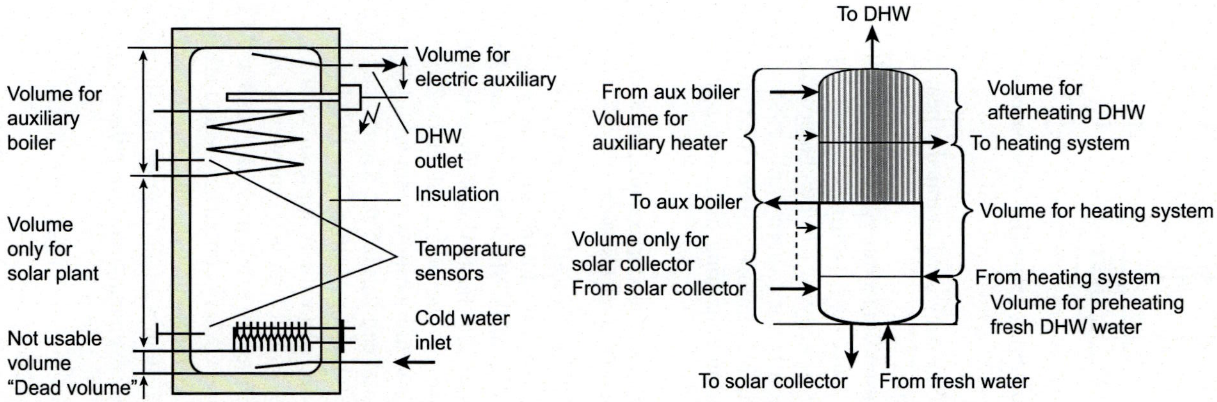

Figure 7.16 Zones for hot water of a DHW system (left) and a solar combisystem (right) (Streicher and Bales, 2005).

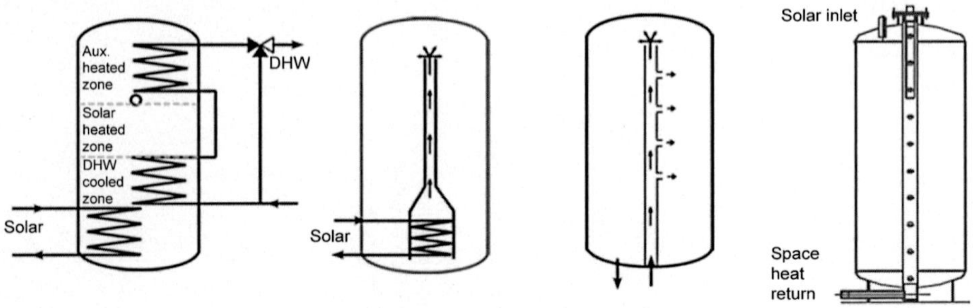

Figure 7.17 Four different stratifying devices for solar water tanks (Streicher and Bales, 2005).

The efficiency of the technology used to transfer heat from the heat sink (nighttime outdoor air) to the building.

The efficiency of a night-cooling strategy lies in the ability of the building inertia to store cool during the night and to use it during the next day. The role of inertia appears then double-linked to this strategy. It can be characterized by means of the storage efficiency and the utilization factor (Allard et al., 1998; Alvarez et al., 1997).

There are several alternatives to incorporate PCM in buildings. Tyagi and Buddhi (2007) present a classification consisting of three types:

- PCMs in building walls
- PCMs in other building components (for example, subfloor or ceiling systems)
- PCMs in separate heat- or cold-storage systems

Athienities and Chen (2000) showed the possibility of using PCMs in underfloor heating systems. The idea is the use of a radiant heating system but with increased performance due to the inclusion of PCMs. The costs have been shown to be reduced if the PCM is used with electrically heated underfloor systems due to the reduction of peak loads and to the use of cheaper night electricity. Similarly, Nagano et al. (2000) presented a floor air-conditioning system with latent heat storage in buildings. Floor size of the experimental cell was 0.5 m^2. Granulated PCM was made of foamed-waste glass beads and a mixture of paraffin. The PCM-packed bed of 3 cm thickness was installed under the floorboard with multiple small holes. The change in room temperature and the amount of stored heat were measured, and results showed the possibilities of cooling load shifting by using packed granulated PCMs.

These new solutions use PCMs in containers, which are located inside an air chamber, in which air is moved by fans with variable flow (forced convection). This air flow is controlled by a control system that can change the air-flow rate, the origin, and the destination of air flow. They even can totally stop the system. The PCM containers are sealed, made of a material that does not react with the PCM contained, and stimulate heat exchange with air around them. In principle, three basic forms have been considered: fins, cylinders, and spheres.

In addition, ceiling boards can incorporate PCM for heating and cooling of buildings. An example is that developed by Kodo and Ibamoto (2002), in which PCMs are used for peak-shaving control of air-conditioning systems in an office building

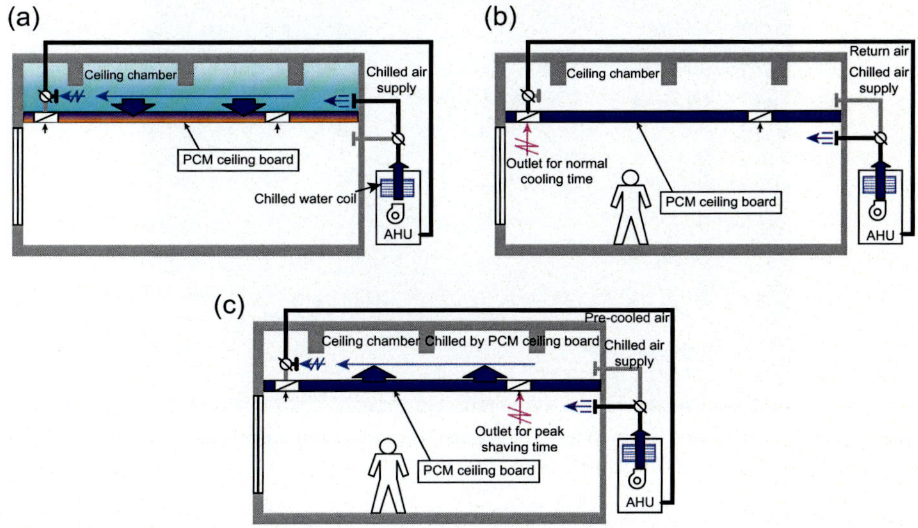

Figure 7.18 Ceiling board with PCM for cooling peak shaving: (a) overnight thermal storage time; (b) normal cooling time; (c) peak-shaving control time (Kodo and Ibamoto, 2002).

(Figure 7.18). The authors claim that these systems have some advantages over conventional building thermal storage systems that use concrete floor slabs:

1. More efficient thermal storage is expected, because high-density cool air pools on the PCM ceiling board that forms the floor of the ceiling space.
2. All of the ceiling board can be used for thermal storage, because the cool air can flow through the ceiling chamber without being interrupted by beams.
3. Because the surface temperature of the ceiling board is kept at the PCM melting point for an extended period, the indoor thermal environment, including the radiant field, can be improved.

Ceiling boards incorporated with PCMs for air-conditioning systems play an effective role on the peak-shaving control. Saman and Belusko (1997) developed a roof-integrated solar air-heating storage system. The latent heat storage unit, in which an existing corrugated iron roof sheet is used as solar collector, is to store heat during the day and supply the heat at night or when sunshine is unavailable. Besides experimental analysis, many numerical works were also carried out on the thermal performance analyses of this system (Vailaltojjar and Saman, 2001; Saman et al., 2005).

Koschenz and Lehmann (2004) put forward a new concept of thermally activated ceiling panel for refurbished buildings. In this system, the mixture of microencapsulated PCM and gypsum was poured into a sheet-steel tray, which was used as a support for maintaining the mechanical stability of the panels. A capillary water tube system was applied to control the thermal mass. They tested the thermal performance of this system and indicated that only a 5 cm layer of microencapsulated PCM and gypsum was enough for a standard office to keep within comfortable temperatures.

Figure 7.19 Experimental setup to test a ventilated façade with macroencapsulated PCMs in its air cavity located in Spain (de Gracia et al., 2013a).

A new type of ventilated façade with macroencapsulated PCM in its air cavity was developed by de Gracia et al. (2013a). The thermal performance of this special building envelope was experimentally tested to analyze its potential in reducing the cooling demand during the summer season (de Gracia et al., 2013a) and the winter season (de Gracia et al., in press) in the Continental Mediterranean climate, and it was numerically extrapolated to different performance scenarios (de Gracia et al., 2013b). Two identical house-like cubicles located in Puigverd de Lleida (Spain) were monitored during summer 2012, and in one of them, a ventilated façade with PCM was located in the south wall (Figure 7.19). Six automatized gates were installed at the different openings of the channel to control the operational mode of the façade. This versatility allows the system to be used as a cold storage unit, as an overheating protection system, or as a night free-cooling application. The experimental results demonstrated the high potential of the night free-cooling effect in reducing the cooling loads of a building. This operation mode could inject air at a temperature below the set point under both severe and mild summer conditions (34.9 MJ day^{-1} and 42.8 MJ day^{-1}, respectively). The system can successfully prevent the overheating effect between the PCM solidification and melting periods, because the air inside the cavity was even lower than the outer environmental temperature during the peak load. The thermal performance of this system is very sensitive to the weather conditions and the cooling demand of the final users.

7.2.2.3 Sorption systems

Only a few applications using liquid absorption can be found. Ruiter (1987) presented an absortion TES cycle using H_2O/NH_3. An experimental setup was built with a net heat output of 5 kW and a storage capacity of 40 kWh. The energy density reached was 111.1 kWh kg^{-1}, based on the total mass of absorbate and weak solution.

Figure 7.20 Double-stage $NaOH/H_2O$ prototype built at the Swiss Federal Laboratories for Materials Science and Technology (EMPA) (Weber and Dorer, 2008).

The pair $NaOH/H_2O$ was studied by Weber and Dorer (2008) using a single-stage closed absorption prototype for long-term heat storage. The prototype built has three storage tanks for water, strong solution, and weak solution (Figure 7.20).

Based on a three-phase absorption cycle, ClimateWell (Jonsson et al., 2000; Bolin, 2005; Olsson and Bolin, 2007) developed a system combining short-term absorption thermal storage and solar cooling technologies. The tests carried out showed that with a 35 kWh heat input, a cooling storage capacity of 22 kWh could be obtained. The calculated energy density for LiCl was 253 kWh m^3, giving a final energy density 1.2 times that of water.

Quinnell et al. (2011) and Quinnell and Davidson (2012) presented a concept using a single storage vessel for storing liquid calcium chloride in a closed liquid absorption system. The storage tank in this prototype is designed to provide higher energy density and to decrease thermal losses during the storage process (Figure 7.21).

A demonstrative $LiBr/H_2O$ prototype with a heat storage capacity of 8 kWh and a discharge rate of 1 kW has recently been developed by N'Tsoukpoe et al. (2013) (Figure 7.22). The prototype is based on a long-term absorption storage cycle and has two storage tanks and a reactor with two vertical falling-film heat exchangers. The achieved charging power was 2–5 kW and the heat storage up to 13 kWh. As in previous prototypes, crystallization of the salts was the main problem found.

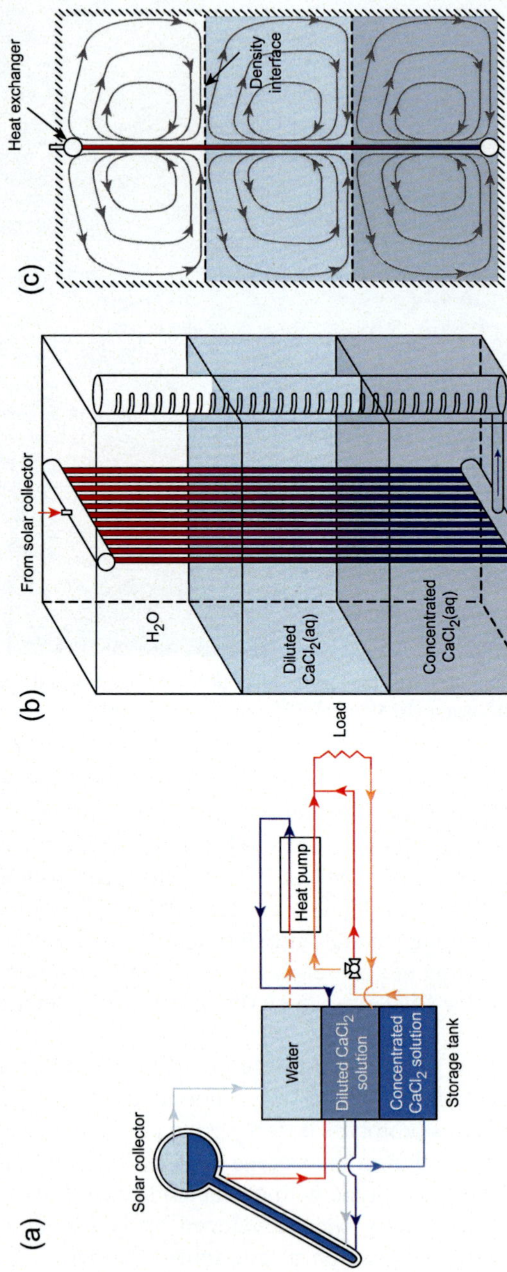

Figure 7.21 Closed $CaCl_2/H_2O$ absorption heating system with a single-store vessel: (a) system schematic (Quinnell and Davidson, 2012); (b) storage tank schematic (Quinnell et al., 2011); (c) anticipated convective flow patterns during sensible charging (Quinnell et al., 2011).

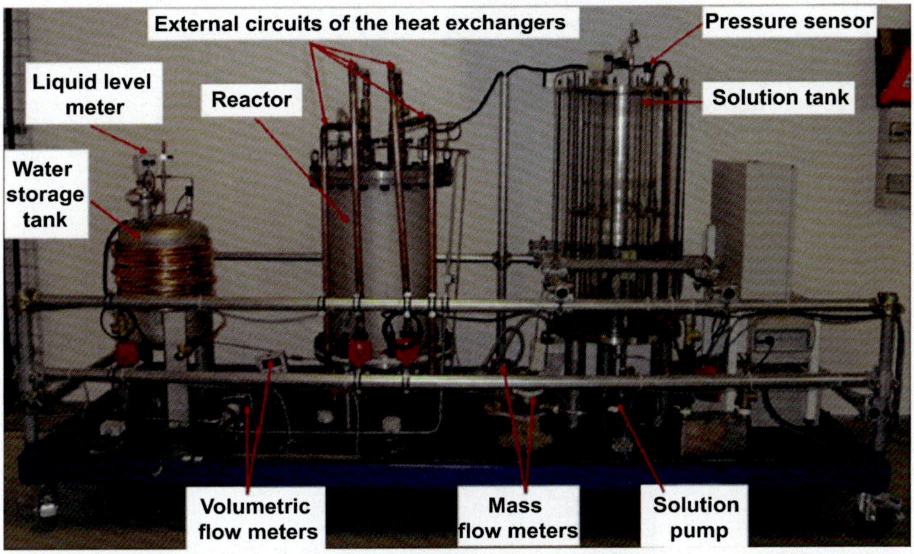

Figure 7.22 LiBr/H_2O absorption prototype developed by N'Tsoukpoe et al. (2013).

The prototype MODESTORE used the positive points of the pair silica gel/H_2O, such as being environmentally harmless and relatively cheap and having low desorption temperatures (Figure 7.23). This prototype used a spiral heat exchanger containing 200 kg of silica gel (Jaenig et al., 2006). The storage capacity obtained in the laboratory was 13 kW. The authors concluded that silica gel is not suitable for long-term sorption TES.

Lu et al. (2003) used a closed-adsorption cold-storage system with the working pair zeolite 13X/water. This prototype has one adsorber and a cold-storage tank (Figure 7.24). The average cooling power reported is 4.1 kW, and the total experimental capacity of the cold storage 5.5 kWh when the temperature of adsorption bed reached its maximum value of 125 °C.

Boer et al. (2004) developed the SWEAT prototype of a modular chemical adsorption cooling system using the working pair Na_2S/H_2O (Figure 7.25). The module has a shell and tube design, a condenser, and an evaporator coil. The test results showed that a cold storage capacity of 2.1 kWh and a cooling COP of 0.56 were achieved with a heat input of 3.7 kWh.

Mauran et al. (2008) presented a storage prototype using the reversible chemical reaction between $SrBr_2$ and $SrBr_2 \cdot H_2O$. The reactor integrated an evaporator/condenser for the solid—gas reaction. The heating and cooling power obtained proved that the most challenging feature that needs to be overcome in this type of concept is the heat transfer problem encountered.

A solar air-conditioning pilot plant with a daily cooling capacity of 20 kWh with a working pair of $BaCl_2/NH_2$ was presented by Stitou et al. (2011) (Figure 7.26). The heat input is at 60 to 70 °C from 20 m^2 of flat-plate solar collectors. Along with

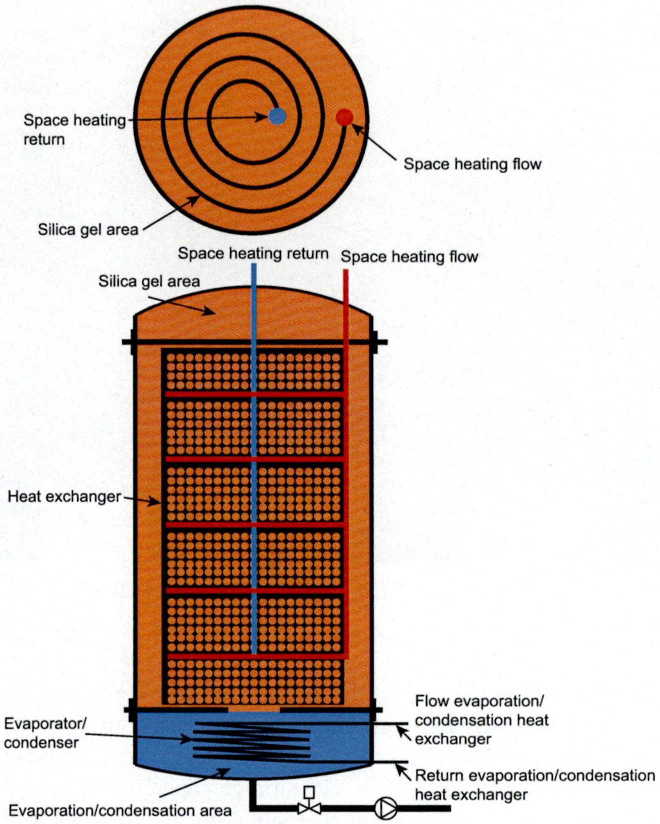

Figure 7.23 MODESTORE prototype (Jaenig et al., 2006).

Mauran et al. (2008), this prototype includes expanded graphite in the design to improve the heat transfer. The prototype also includes a hot PCM tank to store the excess solar heat and a cold PCM tank to supply cooling when the sorption reaction is not available. Experiments during 2 years showed an average yearly efficiency of solar collectors of 0.4 to 0.5 and COP of 0.3 to 0.4. The daily storage capacity was about 0.8−1.2 kWh m^{-2} plate solar collector at 4 °C.

The ZAE Bayern installed a large-scale open-adsorption TES using zeolite 13X/water to heat a school building in winter and to cool a jazz club in summer in Munich (Hauer, 2002, 2007). Figure 7.27 shows the heat flux during charging and discharging. The prototype obtained storage densities of 124 kWh m^{-3} for heating and 100 kWh m^{-3} for cooling with a COP of 0.9 and 0.86, respectively.

Another design is that named chemical heat transfer−low temperature (CWT-NT) (Kerkes et al., 2011; Kerskes et al., 2012; Metter et al., 2013). This concept consists of a long-term thermochemical energy storage integrated in a solar thermal combisystem for a composite of zeolite and salt (Figure 7.28). The system is designed as an open system using ambient air for charging and discharging.

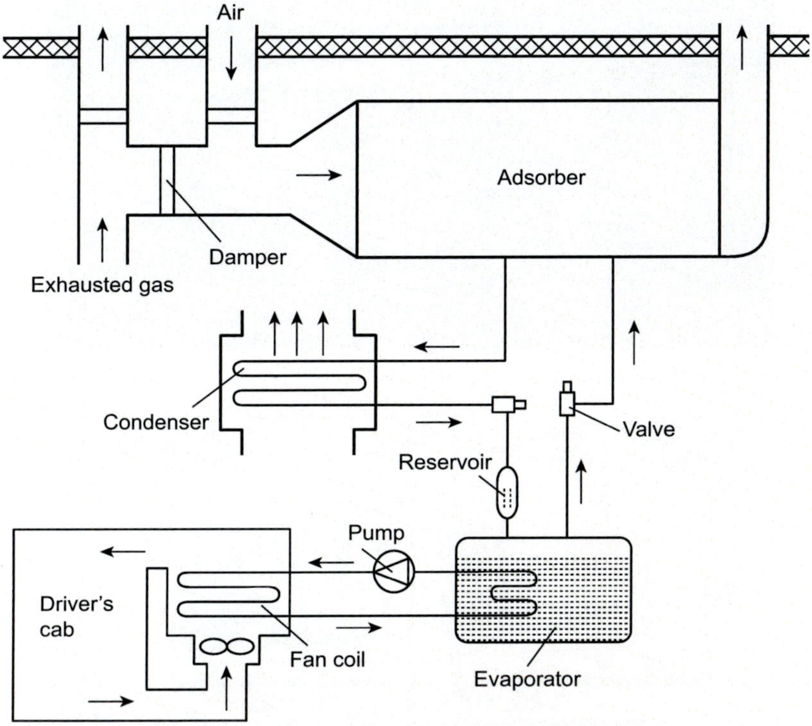

Figure 7.24 Zeolite/water adsorption cold storage system (Lu et al., 2003).

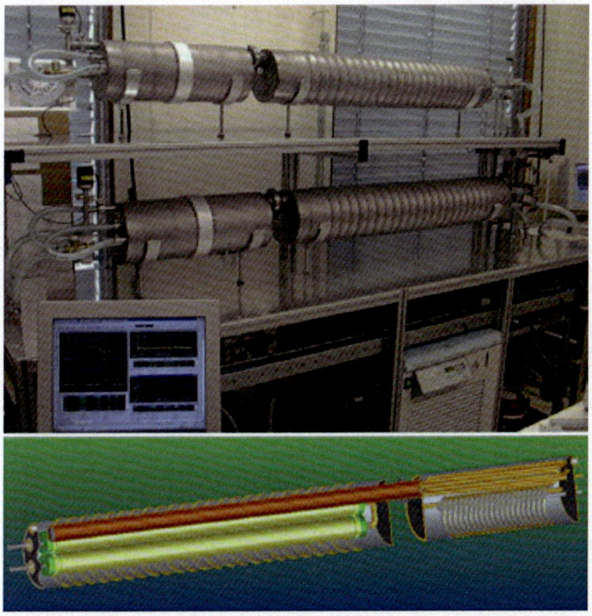

Figure 7.25 SWEAT storage concept (Boer et al., 2004).

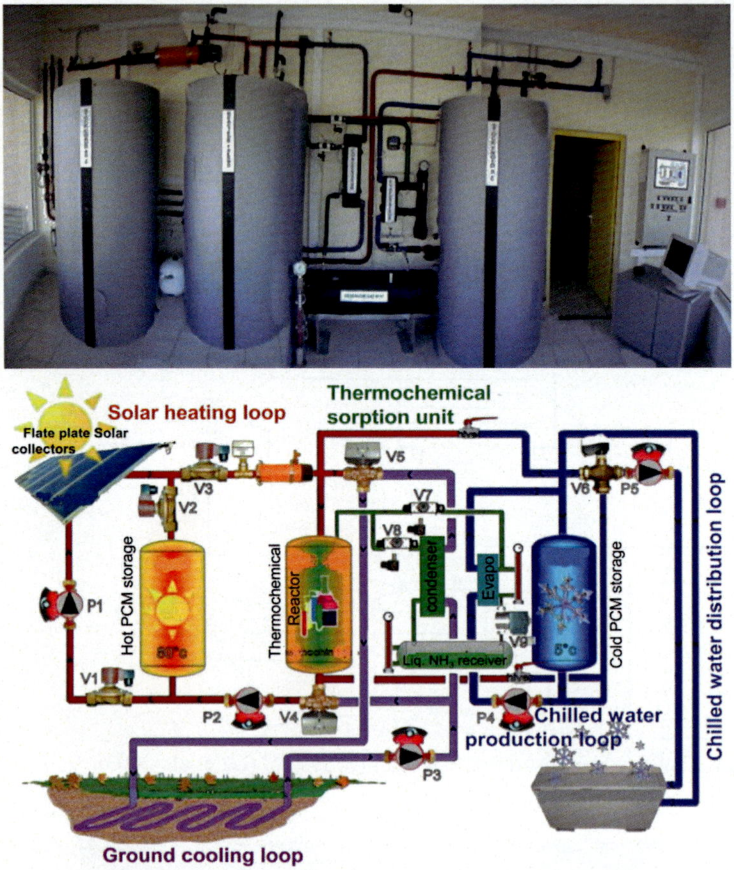

Figure 7.26 Solar sorption pilot plant for air conditioning (Stitou et al., 2011).

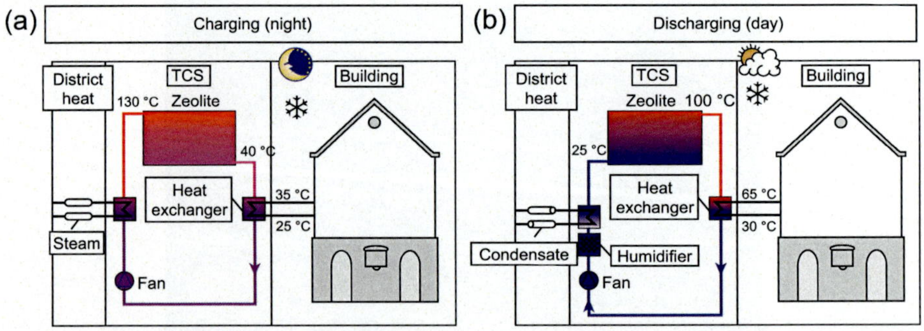

Figure 7.27 Open adsorption TES system connected to the district heating system in Munich (Hauer, 2002).

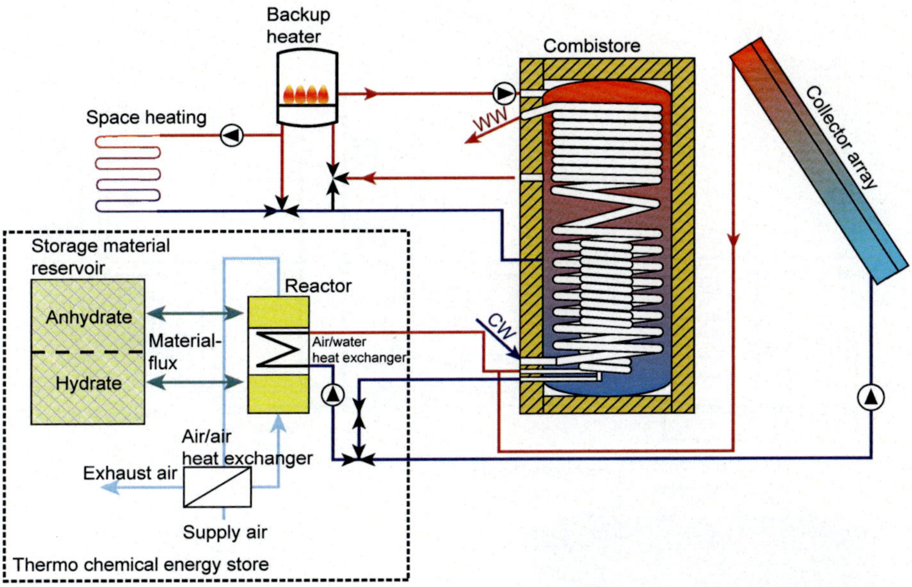

Figure 7.28 CWT-NT concept schematics (Kerskes et al., 2012).

7.2.3 Seasonal storage

7.2.3.1 Underground thermal energy storage

Underground thermal energy storage (UTES) with both boreholes (BTES) and aquifers (ATES) are the most developed storage concepts and are mostly used for seasonal storage. The description of the concepts can be found in Paksoy (2007) and Mehling and Cabeza (2008).

Heat storage in ATES consists in extracting groundwater from a well, heating this water with an available heat source, and then reinjecting it back into the aquifer in another well. The estimated heat storage capacity of 10^5 m^3 of aquifer is 3 MJ for each 10 K temperature range (Hasnain, 1998).

7.2.3.2 Water pits and solar ponds

Novo et al. (2010) reviewed TES systems in large basins (water tanks and gravel–water pits). These authors claimed that the energy costs can be reduced with increasing storage volume in large-scale solar applications.

The most common use of water tanks in Europe is in connection with solar collectors for production of warm water for space heating and/or DHW. The main application is in smaller solar plants such as described above, but there are some examples of large water tanks being used for seasonal storage and also used as a buffer storage in connection with large-scale solar heating systems.

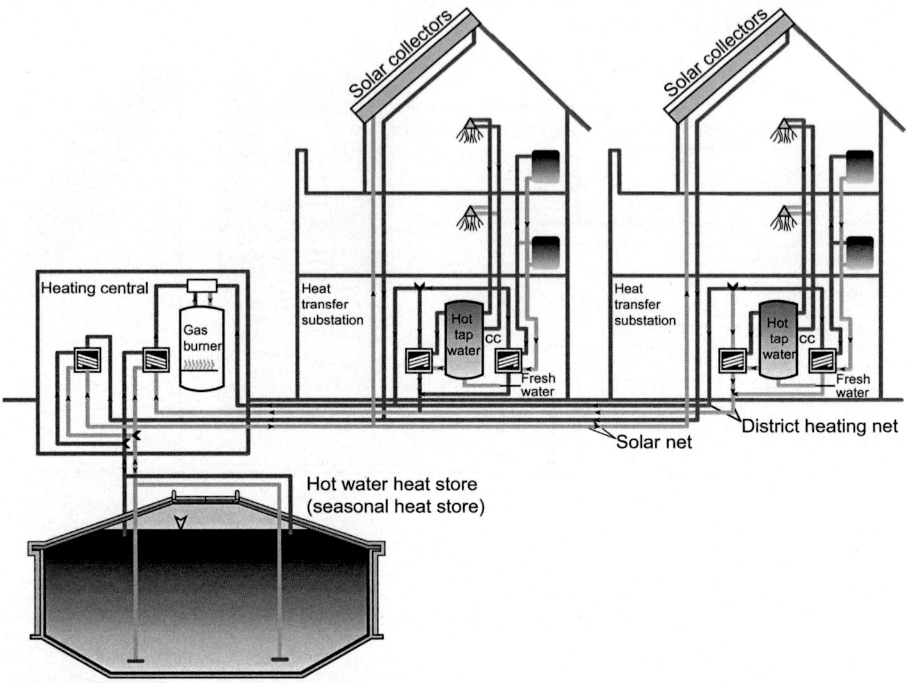

Figure 7.29 Scheme of the system installed in Friedrichshafen, Germany (Schmidt et al., 2004).

These large water tanks usually consist of a reinforced concrete tank partially buried in the ground. It is thermally insulated at least in the roof area and on the vertical walls. Usually steel liners are introduced in the structure to guarantee water tightness and to reduce heat losses caused by vapor transport through the walls (Schmidt et al., 2004). Such a system was installed in Germany (Figure 7.29) together with other concepts.

Storage pits are usually filled with water, but sometimes also rock is added in the pit. Pits are normally buried in the ground and need to be waterproofed and insulated at least at the side walls and on the top. The watertight plastic liner is filled with a gravel—water mixture that constitutes the storage material. Heat is charged into and discharged out of the store either by direct water exchange or by plastic piping installed in different layers inside the store. No other bearing structure is necessary apart from the cover (lid) that could be used for other purposes. The gravel—water mixture has lower specific heat capacity than water alone; for this reason, the volume of the whole basin has to be approximately 50% higher compared to hot-water heat storage to obtain the same heat storage capacity.

Solar ponds are large volumes of saline solution with higher concentration of salts at the bottom than at the top (Figure 7.30). Solar ponds are an economical method to collect and store solar thermal energy in the temperature range 50—95 °C (Novo et al., 2010).

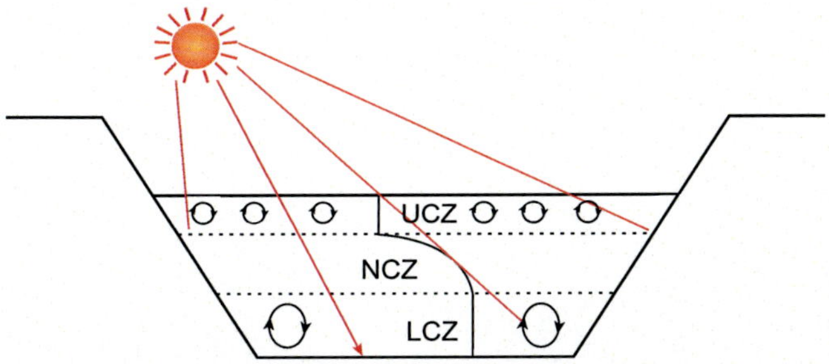

Figure 7.30 Schematic of a solar pond (UCZ = upper convective zone; NCZ = nonconvective zone; LCZ = lower convective zone) (Tatsikjoudoung et al., 2013).

Novo et al. (2010) presented a comparison of these storage systems, stating that the gravel–water pit technology can reduce construction costs and the upper part of the store can be used as part of the residential area but needs more volume to store the same thermal energy than a water tank design. The demonstration sites carried out showed that solar collector efficiency and heat losses from the storage tank and the piping network are most important.

7.2.3.3 Thermochemical storage

Most of the concepts presented in Section 7.2.2.2 as sorption systems are being further developed today for seasonal storage.

7.3 R&D needs and future trends in technological development, markets, and applications

The R&D needs have been captured by the European Technology Platform on Renewable Heating and Cooling (RHC Platform—www.rhc-platform.org) Strategic Research Priorities (Axel et al., 2012). The RHC Platform claims that the emphasis of scientific research, development, and demonstration activities must be focused toward storage technologies that enhance the performance of energy systems and facilitate the integration of renewable energy systems in buildings. Table 7.6 presents the strategic and research priorities for TES according to the RHC Platform.

Because many technologies exist at laboratory scale, the future of TES applications depends on the reduction of costs and improving the ability to efficiently shift energy demand over days, weeks, or seasons. To achieve these objectives, efforts should be focused on advanced sensible heat storage, PCMs, sorption, and thermochemical methods. The most promising areas of research are in latent heat storage and novel thermochemical concepts. Decentralized systems and stores connected to the district heating and cooling networks have potential, so both must be considered.

Table 7.6 Strategic and research priorities for TES according to the RHC platform (Axel et al., 2012)

	Basic research	Applied R&D	Demonstration
Sensible TES	• Materials research for the reduction of heat losses • Materials research for high-temperature storage with high thermal conductivity • Fluids combining heat transfer and heat storage	• Microbiology in underground thermal energy storage (UTES) systems • Flexible volume tank systems • Development of new methods of TES materials' analysis	• Optimization of hydraulics in advanced water stores, reduction of mixing, and increased stratification • Control strategies for integrating sensible stores into the smart grid • High-temperature underground storage (HT-UTES)
Latent TES	• Optimization of phase change heat storage • Fluids combining heat transfer and heat storage	• Integration of phase change materials in building elements	• Software algorithms and codes for ERBP enabling software packages
Thermochemical storage	• Materials for thermochemical heat storage • Fluids combining heat transfer and heat storage	• Optimization of thermochemical heat storage processes • High-temperature thermochemical systems	
Research priorities at system level	• Materials for storage containment	• Advanced sensoring in storage systems • Distributed thermal energy storage for smart electricity grids in smart cities	• Optimized integration of UTES systems • Advanced control strategies • Storage of rejection heat in solar cooling processed and solar power plants • System evaluation
Nontechnological priorities	• Education and training • Knowledge of system performance • Labeling or certification of TES devices • Legal framework UTES (ATES/BTES) • Public awareness		

Another topic is the improvement of the properties of TES materials, especially in improving the lifetime chemical and physical stability. The durability of new systems and their parts also needs to be assessed to ensure their long-term performance.

According to Mahlia et al. (2014), there are three key barriers for the deployment of the energy storage technology, regulations, and utility processes that disfavor energy storage, costs, and lack of awareness of energy storage benefits. These three barriers can also be applied to TES. Important is the fact that there is no experience in deploying energy storage at large scale, so policymakers lack conclusive data about the costs and energy savings capacity of TES.

Sources of further information and advice

Further information on the topic can be found in the published books such as Dincer and Rosen (2002), Hadorn (2005), Paksoy (2007), Mehling and Cabeza (2008), and a chapter in Cabeza (2012).

Many reviews have been published on the topic: Zalba et al. (2003), Farid et al. (2004), Kenisarin and Mahkamov (2007). Sharma et al. (2009), Gil et al. (2010), Cabeza et al. (2011), Cot-Gores et al. (2012), Rathod and Banerjee (2013), and Soares et al. (2013).

Important is the work developed within the International Energy Agency Implementing Agreements. Worth mentioning are Energy Conservation through Energy Storage IA (ECES IA—www.iea-eces.org) and the Solar Heating and Cooling IA (SHC Program—www.iea-shc.org).

References

Ahmad, M., Bontemps, A., Sallée, H., Quenard, D., 2006. Thermal testing and numerical simulation of a prototype cell using light wallboards coupling vacuum isolation panels and phase change material. Energ. Buildings 38, 673–681.

Allard, F., Santamouris, M., Álvarez, S., Guarraccino, G., Maldonado, E., 1998. Natural Ventilation in Buildings: A Design Handbook. Earthscan/James & James, Berlin.

Alvarez, S., Maestre, I., Velazquez, R., 1997. Design methodology and cooling potential of the environmental heat sinks. Int. J. Sol. Energ. 19 (1–3), 179–197.

Athienitis, A.K., Liu, C., Hawes, D., Banu, D., Feldman, D., 1997. Investigation of the thermal performance of a passive solar test-room with wall latent heat storage. Build. Environ. 2 (5), 3405–3410.

Athienities, A., Chen, Y., 2000. The effect of solar radiation on dynamic thermal performance of floor heating systems. Sol. Energ. 69 (3), 229–237.

Axell, M., Bakker, M., Brunialti, A., Cabeza, L.F., Calia, E., Corberan, J.M., de Boer, R., Fevrier, N., Freni, A., Froning, S., Griffiths, P., Hadorn, J.C., Jakob, U., Landolina, S., Lunqvist, P., Monsberer, M., Nordman, R., Nowak, T., Oltersdorf, T., Papillon, P., Py, X., Sanner, B., Schossig, P., Sparber, W., Spoelstra, S., Trigg, L., van Aarssen, M., van Bael, J., van Helden, W., Werner, S., Weiss, W., Ziegler, F., 2012. Strategic research priorities for renewable heating & cooling cross-cutting technology. In: European Technology Platform on Renewable Heating and Cooling.

Barreneche, C., de Gracia, A., Serrano, S., Navarro, M.E., Borreguero, A.M., Fernández, A.I., Carmona, M., Rodríguez, J.F., Cabeza, L.F., 2013. Comparison of three different devices available in Spain to test thermal properties of building materials including phase change materials. Appl. Energ. 109, 421–427.

Boer, R.D., Haije, W.G., Veldhuis, J.B.J., Smeding, S.F., 2004. Solid-sorption cooling with integrated thermal storage: the SWEAT prototype. In: Proceedings of HPC 2004-3rd International Heat Powered Cycles Conference, Larnaca, Cyprus.

Bolin, G., 2005. Chemical Heat Pump Working According to the Hybrid Principle. International patent publication no. WO 2005/054757 A1.

Cabeza, L.F., 2005. Storage techniques with phase change materials. In: Hadorn, J.C. (Ed.), Thermal Energy Storage for Solar and Low Energy Buildings. Universitat de Lleida, Spain, pp. 77–105.

Cabeza, L.F., Ibáñez, M., Solé, C., Roca, J., Nogués, M., 2006. Experimentation with a water tank including a PCM module. Sol. Energ. Mat. Sol. C. 90, 1273–1282.

Cabeza, L.F., Castellón, C., Nogués, M., Medrano, M., Leppers, R., Zubillaga, O., 2007. Use of microencapsulated PCM in concrete walls for energy savings. Energ. Buildings 39 (2), 113–119.

Cabeza, L.F., Castell, A., Barreneche, C., de Gracia, A., Fernández, A.I., 2011. Materials used as PCM in thermal energy storage in buildings: a review. Renew. Sust. Energ. Rev. 15, 1675–1695.

Cabeza, L.F., 2012. Thermal energy storage. In: Sayigh, A. (Ed.), Comprehensive Renewable Energy, vol. 3. Elsevier, Oxford, pp. 211–253.

Castell, A., Martorell, I., Medrano, M., Pérez, G., Cabeza, L.F., 2010. Experimental study of using PCM in brick constructive solutions for passive cooling. Energ. Buildings 42, 534–540.

Castellón, C., Castell, A., Medrano, M., Martorell, I., Cabeza, L.F., 2009. Experimental study of PCM inclusion in different building envelopes. J. Solar Energy Eng, Trans ASME 131 (4), 0410061–0410066.

Cot-Gorés, J., Castell, A., Cabeza, L.F., 2012. Thermochemical energy storage and conversion: a-state-of-the-art review of the experimental research under practical conditions. Renew. Sust. Energ. Rev. 16, 5207–5224.

Dincer, I., Rosen, M.A., 2002. Thermal energy storage (TES) methods. In: Dincer, I., Rosen, M.A. (Eds.), Thermal Energy Storage: Systems and Applications. John Wiley & Sons, New York, NY, pp. 93–212.

Farid, M.M., Khudhair, A.M., Razack, S.A.K., Al-Hallaj, S., 2004. A review on phase change energy storage: materials and applications. Energ. Convers. Manage. 45, 1597–1615.

de Gracia, A., Navarro, L., Castell, A., Ruiz-Pardo, Á., Álvarez, S., Cabeza, L.F., 2013a. Thermal analysis of a ventilated facade with PCM for cooling applications. Energ. Buildings 65, 508–515.

de Gracia, A., Navarro, L., Castell, A., Cabeza, L.F., 2013b. Numerical study on the thermal performance of a ventilated facade with PCM. Appl. Therm. Eng. 61 (2), 372–380.

de Gracia, A., Navarro, L., Castell, A., Ruiz-Pardo, A., Alvárez, S., Cabeza, L.F., 2013. Experimental study of a ventilated facade with during winter period. Energ. Buildings 58, 324–332.

Gil, A., Medrano, M., Martorell, I., et al., 2010. State of the art on high temperature thermal energy storage for power generation. Part 1—concepts, materials and modellization. Renew. Sust. Energ. Rev. 14, 31–55.

Hadorn, J.C. (Ed.), 2005. Thermal Energy Storage for Solar and Low Energy Buildings. Universitat de Lleida, Spain, pp. 77–105.

Hasnain, S.M., 1998. Review on sustainable thermal energy storage technologies, Part I: heat storage materials and techniques. Energ. Convers. Manage. 39, 1127–1138.

Hauer, A., 2002. Thermal energy storage with zeolite for heating and cooling applications. In: Proceedings of ISHPC 2002-International Sorption Heat Pump Conference. Shanghai, China.

Hauer, A., 2007. Sorption theory for thermal energy storage. In: Paksoy, H.O. (Ed.), Thermal Energy Storage for Sustainable Energy Consumption: Fundamentals, Case Studies and Design, NATO Sciences Series, II. Mathematics, Physics and Chemistry, vol. 234. Springer, Dordrecht, pp. 393–408.

Jaehnig, D., Hausner, R., Wagner, W., Isaksson, C., 2006. Thermo-chemical storage for solar space heating in single-family house. In: Proceedings of 10th International Conference on Thermal Energy Storage, New Jersey, USA.

Jonsson, S., Solsson, R., Kaarebing-Olsson, M., 2000. A Chemical Heat Pump. International patent publication no. WO 00/37864 A1.

Kato, Y., 2007. Chemical energy conversion technologies for efficient energy use. In: Paksoy, H.O. (Ed.), Thermal Energy Storage for Sustainable Energy Consumption: Fundamentals, Case Studies and Design, NATO Sciences Series, II. Mathematics, Physics and Chemistry, vol. 234. Springer, Dordrecht, pp. 377–391.

Kenisarin, M., Mahkamov, K., 2007. Solar energy storage using phase change materials. Renew. Sust. Energ. Rev. 11, 1913–1965.

Kerskes, H., Mette, B., Bertsch, F., Asenbeck, S., Drück, H., 2011. Development of a thermo-chemical energy storage for solar thermal applications. In: Proceedings of ISES Solar World Congress 2011. Kassel, Germany.

Kerskes, H., Mette, B., Bertsch, F., Asenbeck, S., Drück, H., 2012. Chemical energy storage using reversible solid/gas-reactions (CWS)-results of the research project. Energy Procedia. 30, 294–304.

Kodo, T., Ibamoto, T., October 1–2, 2002. Research on using the PCM for ceiling board. In: IEA ECES IA, Annex 17, 3rd workshop, Tokyo, Japan.

Koschenz, M., Lehmann, B., 2004. Development of a thermally activated ceiling panel with PCM for application in lightweight and retrofitted buildings. Energ. Buildings 36, 567–578.

Kousksou, T., Bruel, P., Cherreau, G., Leoussoff, V., El Rhafiki, T., 2011. PCM storage for solar DHW: from an unfulfilled promise to a real benefit. Sol. Energy 85, 2033–2040.

Kuznik, F., Virgone, J., Noel, J., 2008a. Optimization of a phase change material wallboard for building use. Appl. Therm. Eng. 28, 1291–1298.

Kuznik, F., Virgone, J., Roux, J.J., 2008b. Energetic efficiency of room wall containing PCM wallboard: a full-scale experimental investigation. Energ. Buildings 40, 148–156.

Kuznik, F., Virgone, J., 2009. Experimental investigation of wallboard containing phase change material: data for validation of numerical modelling. Energ. Buildings 41, 561–570.

Kuznik, F., Virgone, J., Johannes, K., 2011. In-situ study of thermal comfort enhancement in a renovated building equipped with phase change material wallboard. Renew. Energ 36, 1458–1462.

Lv, S.L., Zhu, N., Feng, G.H., 2006. Impact of phase change wall room on indoor thermal environment in winter. Energ. Buildings 38, 18–24.

Lu, Y.Z., Wang, R.Z., Zhang, M., Jiangzhou, S., 2003. Adsorption cold storage system with zeolite-water working pair used for locomotive air conditioning. Energ. Convers. Manage. 44, 1733–1743.

Mahlia, T.M.I., Saktisahdan, T.J., Jannifar, A., Hasan, M.H., Matseelar, H.S.C., 2014. A review of available methods and development on energy storage; technology update. Renew. Sust. Energ. Rev. 33, 532–545.

Mandilaras, I., Stamatiadou, M., Katsourinis, D., Zannis, G., Founti, M., 2013. Experimental thermal characterization of a Mediterranean residential building with PCM gypsum board walls. Build. Environ. 61, 93–103.

Mauran, S., Lahmidi, H., Goetz, V., 2008. Solar heating and cooling by a thermochemical process first experiments of a prototype storing 60 kWh by a solid/gas reaction. Sol. Energ. 82, 623–636.

Mehling, H., Cabeza, L.F., 2008. Heat and Cold Storage with PCM: An Up to Date Introduction into Basics and Applications. Springer, Heidelberg, Berlin.

Mette, B., Kerskes, H., Drück, H., Müller-Steinhagen, H., 2013. New highly efficient regeneration process for thermochemical energy storage. Appl. Energ. 109, 352–359.

Nagano, K., Mochida, T., Iwata, K., Hiroyoshi, H., Domanski, R., Rebow, M., 2000. Development of new PCM for TES of the cooling system, Terrastock. In: Benner, M., Hahne, E.W.P. (Eds.), 8th International Conference on Thermal Energy Storage, pp. 345–350.

Neeper, D.A., 2000. Thermal dynamics of wallboard with latent heat storage. Sol. Energ. 68 (5), 393–403.

Novo, A.V., Bayon, J.R., Castro-Fresno, D., Rodriguez-Hernandez, J., 2010. Review of seasonal heat storage in large basins: water tanks and gravel-water pits. Appl. Energ. 87, 390–397.

N'Tsoukpoe, K.E., Le Pierrès, N., Luo, L., 2013. Experimentation of a LiBr-H_2O absorption process for long-term solar thermal storage: prototype design and first results. Energy 53, 179–198.

Olsson, R., Bolin, G., 2007. Chemical Heat Pump Working According to the Hybrid Principle. International patent publication no. WO 2007/139476 A1.

Paksoy, H.O., 2007. In: Thermal Energy Storage for Sustainable Energy Consumption: Fundamentals, Case Studies and Design. NATO Sciences Series, II. Mathematics, Physics and Chemistry, vol. 234. Springer, Dordrecht, pp. 393–408.

Pasupathy, A., Velraj, R., 2008. Effect of double layer phase change material in building roof for year round thermal management. Energ. Buildings 40, 193–203.

Quinnell, J.A., Davidson, J.H., 2012. Mass transfer during sensible charging of a hybrid absorption/sensible storage tank. Energy Procedia. 30, 353–361.

Quinnell, J.A., Davidson, J.H., Burch, J., 2011. Liquid calcium chloride solar storage: concept and analysis. J. Sol. Energy. Eng. 133, 011010–011011-8.

Rathod, M.K., Banerjee, J., 2013. Thermal stability of phase change materials used in latent heat energy storage systems: a review. Renew. Sust. Energ. Rev. 18, 246–258.

Ruiter, J.P., 1987. Storage of Thermal Enregy by Means of an Absorption Cycle (Ph.D. thesis). Delft University of Technology, Delft.

Saman, W.Y., Belusko, M., 1997. Roof integrated unglazed transpired solar air heater. In: Lee, T. (Ed.), Proceedings of the 1997 Australian and New Zealand Solar Energy Society. Paper 66, Canberra, Australia.

Saman, W., Bruno, F., Halawa, E., 2005. Thermal performance of PCM thermal storage unit for a roof integrated solar heating system. Sol. Energ. 78, 341–349.

Schmidt, T., Mangold, D., Müller-Steinhagen, H., 2004. Central solar heating plants with seasonal storage in Germany. Sol. Energ. 76, 165–174.

Schossig, P., Henning, H.-M., Gschwander, S., Haussmann, T., 2005. Micro-encapsulated phase-change materials integrated into construction materials. Sol. Energ. Mat. Sol. C. 89, 297–306.

Sharma, A., Tyagi, V.V., Chen, C.R., Buddhi, D., 2009. Review on thermal energy storage with phase change materials and applications. Renew. Sust. Energ. Rev. 13, 318–345.

Soares, N., Costa, J.J., Gaspar, A.R., Santos, P., 2013. Review of passive PCM latent thermal energy storage systems towards buildings' energy efficiency. Energ. Buildings 59, 82–103.

Stitou, D., Mazet, N., Mauran, S., 2011. Experimental investigation of a solid/gas thermochemical storage process for solar air-conditioning. Energy 41, 261.

Streicher, W., Bales, C., 2005. Combistores. In: Hadorn, J.C. (Ed.), Thermal Energy Storage for Solar and Low Energy Buildings. Universitat de Lleida, Spain, pp. 77–105.

Talmatsky, E., Kribus, A., 2008. PCM storage for solar DHW: an unfulfilled promise? Sol. Energ. 82, 861–869.

Tatsidjodoung, P., Le Pierrès, N., Luo, L., 2013. A review of potential materials for thermal energy storage in building applications. Renew. Sust. Energ. Rev. 18, 327–349.

Tyagi, V.V., Buddhi, D., August 2007. PCM thermal storage in buildings: a state of art. Renew. Sust. Energ. Rev. 11 (6), 1146–1166.

Vakilaltojjar, S.M., Saman, W., 2001. Analysis and modeling of a phase change storage system for air conditioning applications. Appl. Therm. Eng. 21, 249–263.

Weber, R., Dorer, V., 2008. Long-term heat storage with NaOH. Vacuum 82, 708–716.

Xu, X., Zhang, Y.P., Lin, K.P., Di, H.F., Yang, R., 2005. Modeling and simulation on the thermal performance of shape-stabilized phase change material floor used in passive solar buildings. Energ. Buildings 37, 1084–1091.

Yu, N., Wang, R.Z., Wang, L.W., 2013. Sorption thermal storage for solar energy. Prog. Energ. Combust. 39, 489–514.

Zalba, B., Marín, J.M., Cabeza, L.F., Mehling, H., 2003. Review on thermal energy storage with phase change: materials, heat transfer analysis and applications. Appl. Therm. Eng. 23, 251–283.

Hybrid systems for renewable heating and cooling

8

Wolfram Sparber, Roberto Fedrizzi
EURAC Research, Bolzano, Italy

8.1 Introduction

A "**hybrid system**" is a system combining two or more energy sources to provide heating, cooling, and hot water to buildings or industrial processes. Today, renewable sources can augment existing fossil systems or newly installed renewable systems can be augmented by nonrenewable sources. Both qualify as hybrid systems.

In the long term, for optimal sustainability, hybrid systems should be based only on the combination of different renewable energy sources. Already now, several options exist to cover 100% of the demand for heating and cooling in a building from renewable sources.

Hybrid systems are today mainly applied in the **heating sector of buildings**. Within the cooling sector they represent only a niche application.

The application with the largest diffusion is solar thermal systems for domestic hot water and eventually as well as space heating combined with fossil fuel burners (as shown in Figure 8.1). But solar thermal collectors can be part of many different kinds of hybrid energy systems for heating, domestic hot water, and eventually cooling (see, for example, Figure 8.2).

Within the **district heating and cooling sector** many systems—especially for large-scale applications—are not based on one single heating technology, but on several, which can be located geographically in different locations. In these cases, the heat

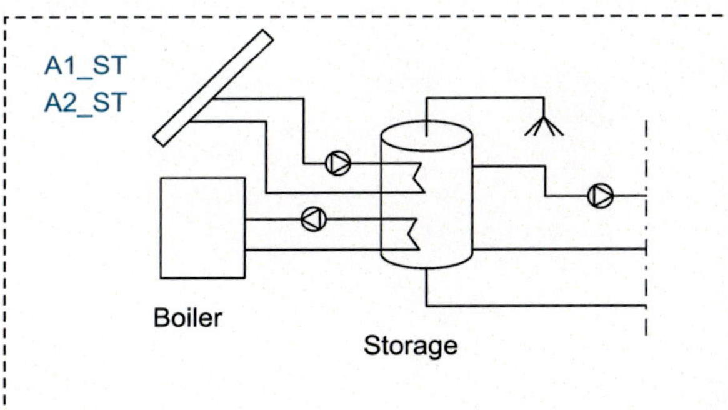

Figure 8.1 Schematic view of a hybrid system for buildings consisting of a solar thermal collector field, a boiler, and storage serving heating and domestic hot water preparation. Source: EURAC.

Renewable Heating and Cooling. http://dx.doi.org/10.1016/B978-1-78242-213-6.00008-4

Figure 8.2 Solar thermal collector field on the green roof of a low-energy building located in the Italian Alps. The collector field is part of a demo hybrid system including a biomass boiler and a thermally driven chiller [1].

sources can be manifold, such as fossil fuels (natural gas, oil, or coal), waste heat applications (waste heat from industrial processes, waste incineration, or cogeneration), and renewable sources (biomass, biofuels, geothermal, solar thermal, or heat pump applications). A schematic view and an example are shown in Figures 8.3 and 8.4.

Hybrid systems have the advantage that the renewable source can be used when available (from a timely point of view) and the single technology can be used in an optimum range of applications (considering, e.g., temperature range). In fact, in the

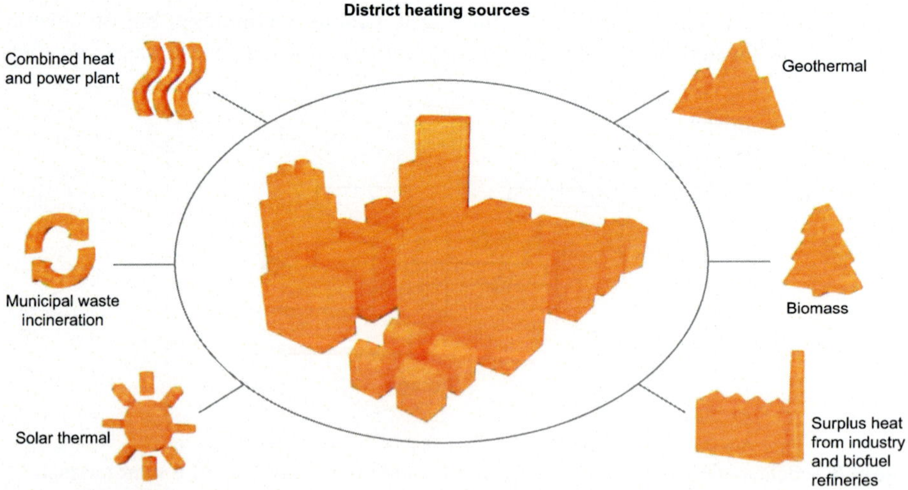

Figure 8.3 Schematic view of different energy sources integrated in a district heating and cooling system [2].

Figure 8.4 Example of a district heating system located in the Italian Alps including solid biomass boilers and natural gas cogeneration units [3].

situations in which the renewable energy source is limited or its application inefficient, the secondary of fossil source works as backup.

The disadvantage is the need for two different technologies and the need to find the most appropriate way to combine and use the single sources. This aspect increases the complexity of the systems (in comparison to single-technology applications) and requires an intelligent way of combining the two sources (hydraulically and regarding the control unit). This aspect leads to higher upfront investment cost (in comparison to single-technology applications), also increasing the complexity of the systems. Therefore, the control unit and a proper adjustment to the way of application can play relevant roles considering the overall efficiency and applicability of the hybrid system.

8.2 Applications of hybrid systems integrating renewable heating and cooling worldwide

Hybrid systems are the combination of two single heating or cooling technologies in one heating and cooling system.

Hybrid systems can be applied in most kinds of heating and cooling applications, ranging from single-family houses to large-scale industrial applications, combined heat and power plants, and district heating and cooling applications.

To allow exploitation of renewable energy sources available on site, hybrid systems are increasingly applied. Hybrid systems can be new systems planned as such from the very beginning or they can be the result of a refurbishment of an existing fossil fuel-based system with which renewable heating and cooling technologies are integrated.

The most commonly applied technology combination up to now is the utilization of a solar thermal collector field for domestic hot water and eventual space heating support combined with natural gas or oil burners for residential buildings. An application that is expected to increase market share is the combination of relatively small-sized air-source heat pumps supporting existing fossil fuel burners.

Considering large-scale applications like district heating systems, biomass burners, deep geothermal applications, solar thermal fields, and/or heat pumps can be integrated in existing systems. Such integration can be structured stepwise, enhancing over the years the capacity of installed renewable heat capacity and reducing the utilization of fossil fuels.

Market figures are usually collected for single technologies alone; it is very rare that specific hybrid system market figures are collected. Nevertheless, to give the reader an impression of the development of hybrid systems in recent years, the market developments of single technologies in different geographical areas suitable for hybrid systems are shown in the following. Next to the market figures, the main technological combinations are described to give the reader an introduction to the topic. For further information on specific technologies and systems, links to websites and books are cited.

8.2.1 Hybrid systems for domestic hot water preparation and heating

Solar thermal systems are a technology that in many cases (especially in developed countries) is not used as a standalone technology but as a part of a hybrid system. In such a hybrid system, the solar thermal collector field is usually combined with a fossil fuel burner (see Figure 8.1). Alternatives are combinations with biomass burners or heat pumps.

In Figure 8.5 the worldwide market development of installed glazed solar thermal capacity is shown. As seen here, since the year 2000 the solar thermal market has

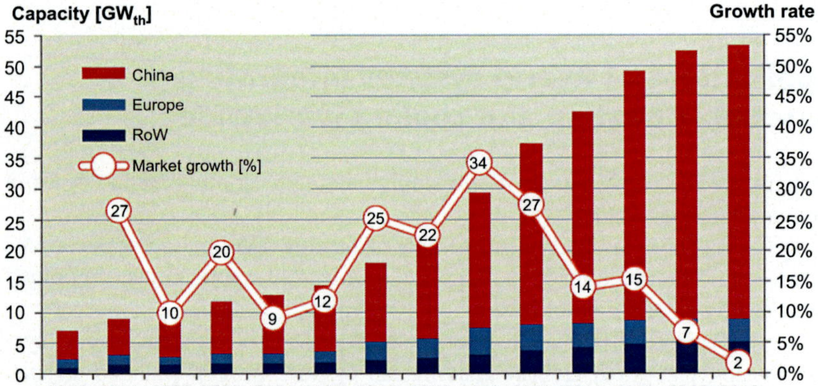

Europe: Eu 28, Albania, Macedonia, Norway, Russia, Switzerland, Turkey
Rest of World (RoW): Asia (India, Japan, South Korea, Taiwan, Thailand), Australia, Canada, Latin America (Barbados, Brazil, Chile, Mexico, Uruguay), New Zealand, MENA Region (Israel, Jordan, Lebanon, Morocco, Palestinian Territories, Tunisia), Sub-Saharan Africa (Mauritius, Mozambique, Namibia, South Africa, Zimbabwe), USA

Figure 8.5 Global market development of glazed solar thermal water collectors [4].

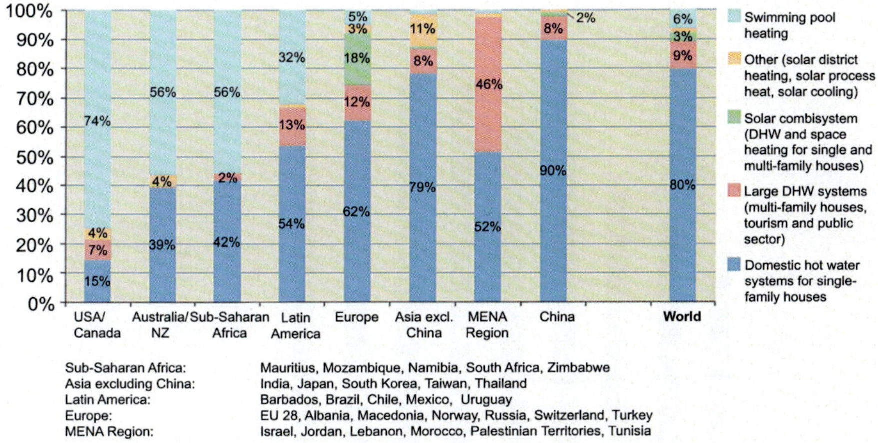

Figure 8.6 Distribution of solar thermal systems by application for the total installed water collector capacity by economic region in operation by the end of 2013 [3].

experienced growth, mainly driven by the Chinese market, which is largely dominating worldwide solar thermal applications.

Considering the kind of application of solar thermal systems, Figure 8.6 shows the domination of domestic hot water systems for single-family houses. This kind of application is responsible for 80% of the total worldwide installed solar thermal capacity. Considering only 2013 newly installed capacity, the large domestic hot water systems—which are always hybrid systems—could increase their share to 17%, whereas the share of swimming pool applications, which are often standalone systems, decreased to 3%. The new installations of 2013 were still dominated by domestic hot water applications for single-family houses, with a share of 77%.

A further technology that is particularly feasible for applications in hybrid systems are **heat pumps**. Regarding heating applications, these might be electrically driven heat pumps combined with solar thermal collectors, heat pumps combined with fossil fuel burners in existing energy systems, or large-scale electrically driven heat pumps for heating support in industry or district heating applications. Regarding cooling applications these might be heat pumps thermally driven by waste heat, district heat, or solar thermal collectors and combined with electrically driven chillers.

The European market development for heat pumps is shown in Figure 8.7. As can be seen strong market growth occurred in the years to 2008. With the worldwide economic crisis, the European building sector was strongly influenced, leading as well to a stabilization of the heat pump market from 2008 to 2013. The dominating technologies within the heat pump market are air–air and air–water heating applications. With regard to hybrid heating systems mainly air–water heat pumps and water–water heat pump systems coupled to geothermal heat exchangers are of interest.

Hybrid systems consisting of **heat pumps and solar thermal systems** have attracted special attention by research and industry in recent years. There are many

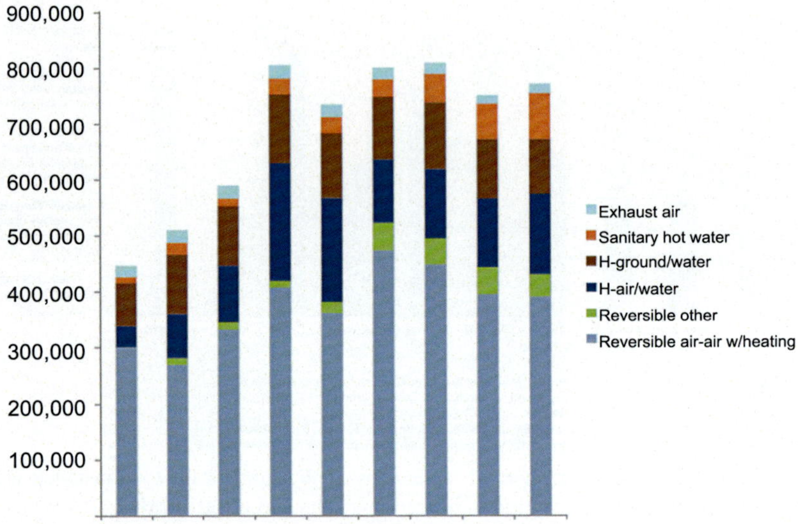

Figure 8.7 Development of heat pump sales in Europe from 2005 to 2013, by category [5].

different ways by which these two technologies can be combined. In Figures 8.8 and 8.9 two examples of hydraulic schemes are shown. The expectation is to reach overall system seasonal performance figures that are significantly higher than the seasonal performance figures of heat pump systems as standalone solutions.

Within the International Energy Agency Solar Heating and Cooling Program, jointly with the Heat Pump Program, an international multi-annual research cooperation project concluded in 2014 [7].

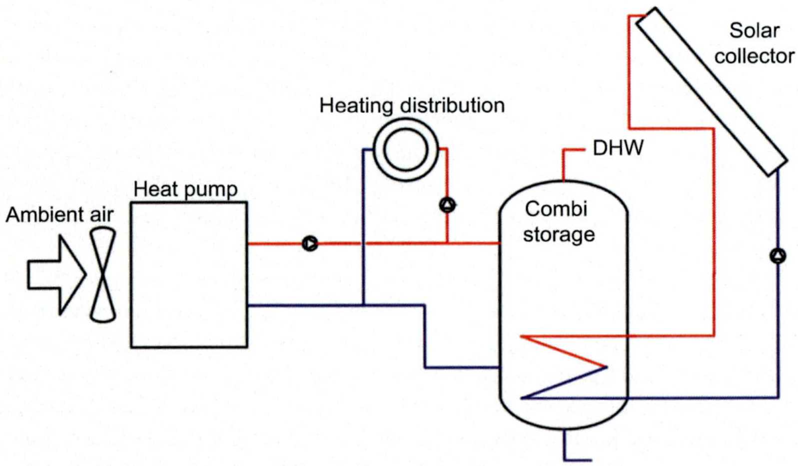

Figure 8.8 Schematic view of a hybrid system consisting of a solar thermal collector field and an air–water heat pump. The two heating technologies are connected in parallel [6].

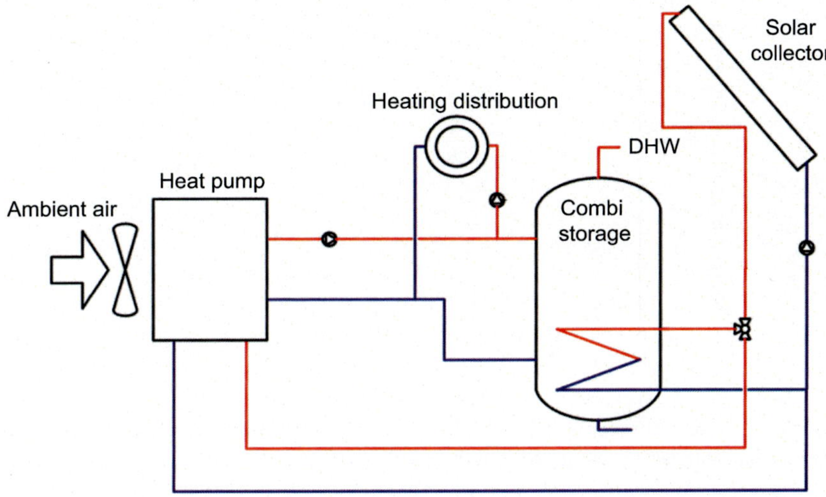

Figure 8.9 Schematic view of a hybrid system consisting of a solar thermal collector field and an air–water heat pump. The two heating technologies are connected in parallel and in serial [6].

Within the project, many different kinds of solar thermal and heat pump systems were analyzed and simulated; specific seasonal performance figures and boundary conditions were defined, and several implemented systems have been monitored in the field for years. It turned out that the theme is much more complex than it might appear just entering the topic. Several of the monitored and simulated systems did not reach expectations. In some cases, because of the enhanced complexity of the systems and non-optimal hydraulics and/or control strategy, the seasonal performance figures remained below those of efficient heat pump systems alone. However, several examples exist in which the design and implementation have been well done, leading to systems with seasonal performance figures clearly exceeding heat pump-only systems.

The most straightforward system designs are systems in which the two sources are connected in parallel. Such systems are directly comparable to hybrid systems combining fossil fuel burners and solar thermal collector fields. But if the fossil fuel burner is replaced by a heat pump, the primary energy consumption and CO_2 emissions can be eventually reduced. The amount of reduction is dependent on the seasonal performance figures of the heat pump and the electrical production mix applied in the region.

In the case of serial connections, the objective is to enhance the performance of the heat pump by means of providing higher temperatures to the heat pump evaporator through the collector field in comparison to an aerothermal source. Detailed analysis and simulations showed that surrounding conditions under which this can be used effectively are rather limited. In the case of geothermal heat source applications, the solar thermal heat can as well be applied in order to regenerate the ground when excess heat is available (usually during summer month).

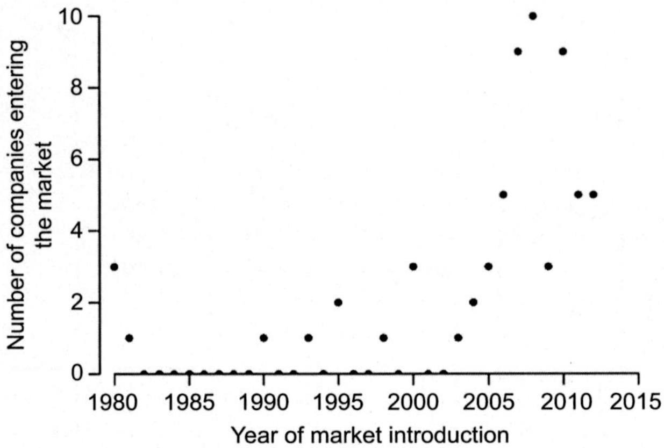

Figure 8.10 Companies entering the market of solar heat pump systems (the oldest system offered by each company is used as the indicator, provided that it is still marketed) [9].

Further information, such as on system design, test procedures, simulation results, and monitoring results of solar and heat pump systems, can be found in the technical handbook [8].

To control the complexity of the system and to assure the correct implementation of the system on site, single companies have introduced prefabricated kit systems for small-scale applications to the market. Such systems in many cases combine in one single kit the heat pump, hot-water storage, the main hydraulic circuit, and the control unit. Thus, the hydraulic work necessary for implementation on site is limited to the connection of the central energy unit to the solar thermal collector field, the ground heat exchanger or outside air unit, and the building heating and domestic hot water circuits. Here, time through installation is limited, and the complexity for the installer is dramatically reduced.

Figure 8.10 shows the number of European companies that, over the years, have introduced prefabricated solar thermal and heat pump systems into the market. The figures are based on a market survey of over 70 companies in Europe offering solar and heat pump systems. As can be seen, a relevant number of systems was introduced only after 2005. Detailed market sales figures for such systems were not available to the authors for the present chapter, but, in general, it can be stated that the application of this technology is still a limited niche market in comparison to the overall solar thermal and heat pump markets.

A system combination that is not a hybrid system in a strict sense, but that can allow a very high degree of renewable energy utilization and is applied and promoted in many countries, is the combination of **heat pumps and solar photovoltaic systems** (see Figure 8.11). In many applications the two energy systems are set up in the same facility but are energetically only indirectly connected. In fact, the photovoltaic (PV) system is connected to the grid, covers the local electricity load if possible, and feeds surplus electricity into the grid. The heat pump in such systems is just one of several different electrical energy loads.

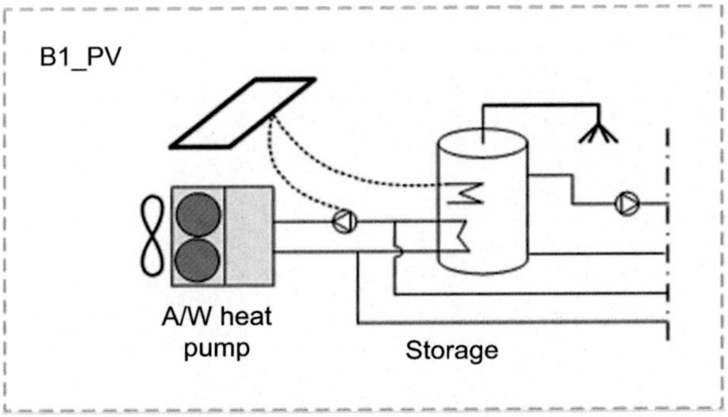

Figure 8.11 Schematic view of a heating system combining an air–water heat pump with a solar photovoltaic system and heat storage.
Source: EURAC.

However, few systems are available on the market that try to match renewable electrical generation with heat pump (HP) electrical consumption. This is obtained by means of:

1. Electric and/or thermal storages that store the PV electricity when available and distribute it to the heat pump when necessary. In this way, self-consumption is maximized.
2. A wise utilization of the thermal loads (demand side management) and a strong interaction with local electrical generation with the electric grid operator. In this way the heat pump can be activated, for example, in advance in times of high PV electrical production and low electrical consumption in the grid.

Most of the heat pump systems described previously are used only for heating and domestic hot water production. In climates where heating is needed in the winter months and cooling is needed in the summer months, **reversible heat pumps** can be applied, which cover both needs and domestic hot water over the entire year. In such systems, the utilization of PV electricity can be very beneficial due to the contemporaneity of solar irradiation availability and the building's cooling load.

8.2.2 Hybrid systems for district heating and cooling applications

Within **district heating systems** in many cases several heating technologies and heat sources are combined with each other, leading to hybrid energy systems. Based on the size of the installed capacity and considering the geographic expansion of district heating systems, no other heating system has such a high flexibility and ability to combine different heat sources and technologies into one single heat distribution system.

Hybrid systems can be the result of combining:

- Waste heat applications (from cogeneration, waste incineration, industry applications) and heating technologies
- Different fuels that are exploited in the system (based on coal, oil, natural gas, biogas, biomass, deep geothermal, solar thermal, heat pumps, etc.)
- Heating units located geographically in different positions of the network exploiting different energy sources

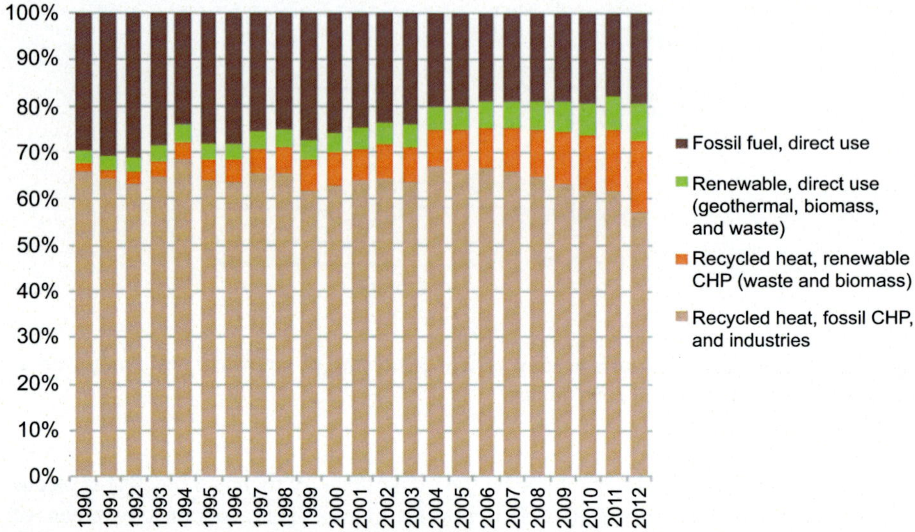

Figure 8.12 Heat sources used to supply the European district heating systems from 1990 to 2012. Source: Data considering 28 EU countries, Prof. S. Werner, Halmstadt University, based on International Energy Agency Energy Balances [10].

Figure 8.12 shows the **applied heat sources** in European district heating systems. As can be seen, the large majority uses recycled heat from cogeneration based on fossil fuels, waste incineration, and renewable sources. To be able to supply the strongly varying heat demand over time, district heating systems combine the recycled heat source with additional heating technologies (mostly fossil or renewable fuels such as solid biomass or biofuels).

In recent years, in several European countries, the added heating capacity strongly moved toward exploiting **renewable energy sources**. In Figure 8.13 the example of Austria is shown. As can be seen in the time frame from 1990 to 2011, oil lost much of its importance whereas natural gas took over its position, keeping the sum of the two fluctuating around 3 TWh of produced heat per year. The overall growth in produced heat was covered mainly by biomass. This led to the 2011 situation in which biomass with over 5 TWh of produced heat became the most important heat source for heat-only production in Austria.

Deep geothermal energy exploitation for district heating and cooling currently represents a niche market within the European heat sources for district heating and cooling. However, the technology is evolving and project experience is growing, so the market is developing with important steps. In Figure 8.14, the deep geothermal systems in operation in 2014 and the ones under development are shown. As can be seen in some of the main European markets, such as France, Germany, Hungary, and Italy, the number of geothermal systems is expected to nearly double by 2018. Many of the district heating and cooling systems exploiting geothermal energy are set up as hybrid systems with fuel burners as an additional heat source.

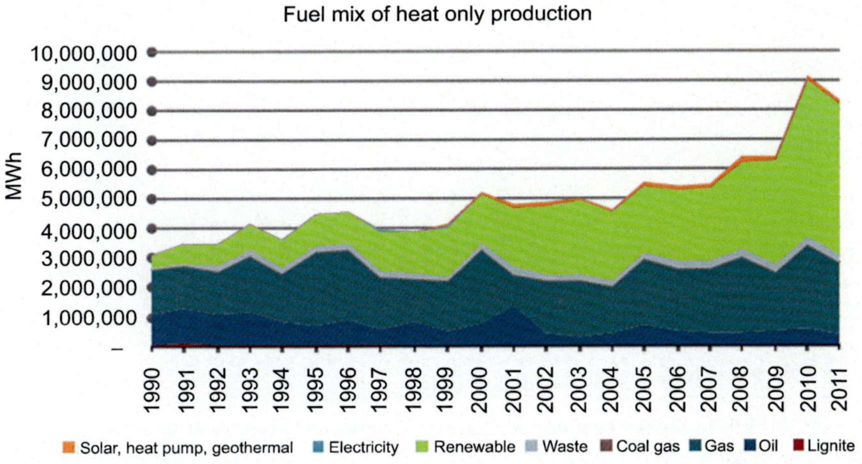

Figure 8.13 Fuel mix of heat-only production in Austrian district heating systems from 1990 to 2011 [10].

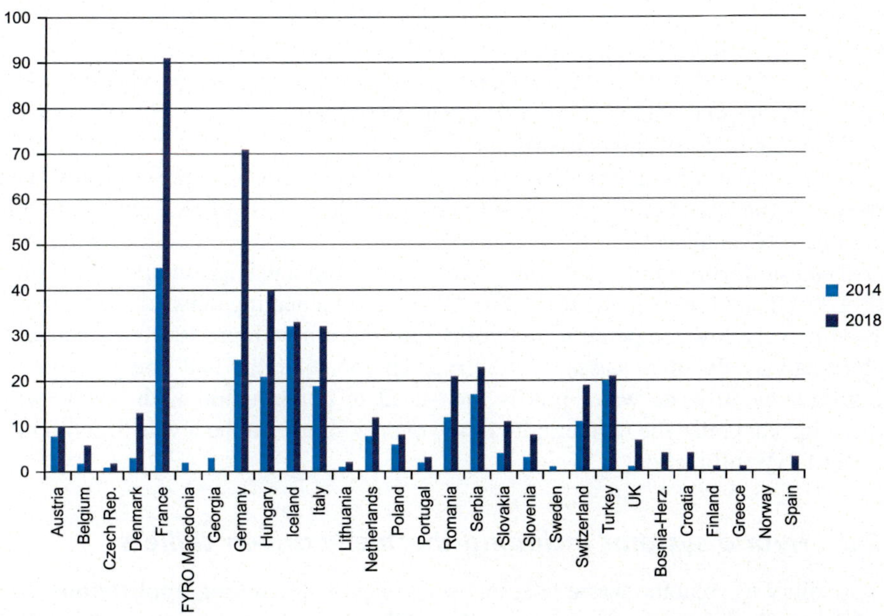

Figure 8.14 District heating systems using deep geothermal energy as heat source in Europe, country by country. Systems under operation in 2014 and under development and expected to be in operation in 2018 [11].

The utilization of **solar energy** in combination with district heating networks has been limited up to now. In Figure 8.15 the development of such systems over recent decades is shown. By the end of 2014, 210 solar thermal systems with a capacity of over 350 kW_{th} were connected to a district heating system [4]. Countries that can be

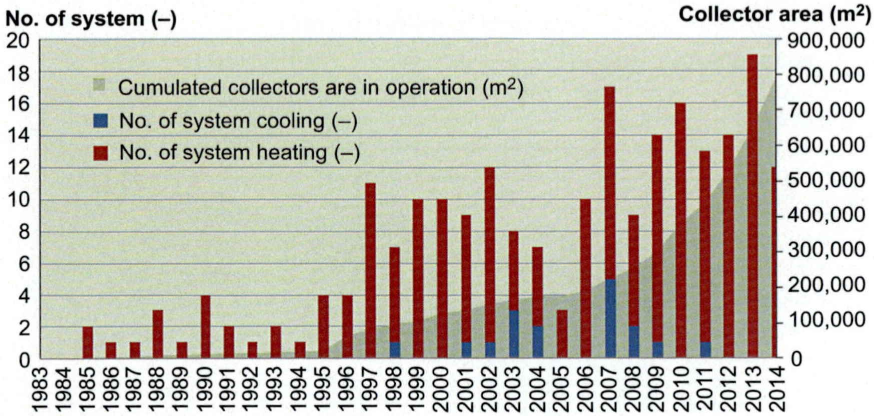

Figure 8.15 Hybrid district heating and cooling systems including solar thermal energy as energy source. Annual newly installed systems and cumulative collector area in operation by the end of 2014.
Source: Prof. J. O. Dalenbeck, Chalmers University of Technology [4].

identified as virtual examples in the sense of the application of this technology are Denmark, Sweden, Austria, or Germany. Here, several systems are encountered in which solar thermal energy is used to integrate partially or to cover a relevant part of the heating loads of the network.

Regarding the solar thermal technology, flat-plate collectors are typically used; however, applications in which concentrating collectors are exploited are in a preliminary phase of analysis.

All these technologies are integrated in traditional high-temperature third- (80−100°C) and fourth-generation (50−60°C) district heating networks. A new generation of very low-temperature networks (ground temperature level) are currently under analysis. In these cases, geothermal and/or solar thermal energy stored in the ground or in large water pits is then used in combination with water−water heat pumps that raise the temperature of the energy source to the levels for both heating and DHW utilization.

8.2.3 Hybrid systems including thermally driven chillers

A possibility to combine waste heat/renewable heat with **cooling applications** is the implementation of thermally driven chillers. These technologies in many cases are combined with an electrically driven heat pump in a hybrid cooling system.

The heat source for thermally driven chillers can be waste heat from cogeneration, waste heat from waste incineration, waste heat from industry processes, or heat from specific renewable heating systems such as solar thermal collector fields. The cooling capacity can be installed centrally, and connection is given via a district cooling system. Another possibility is that the cooling capacity is being installed on site. In these cases as well the heat source can be located on site, or a connection to a centralized heat source can be established through a district heating system. Such applications are increasingly of interest, as

- District heating systems are expanding in many urban areas in Europe and worldwide.
- In many cases heat recovery of cogeneration units are the main heat source for the district heating system and therefore available all year.
- In winter time the heat is directly used for heating and domestic hot water.
- In summer time in many cases the heat is wasted to the environment.
- Thermally driven cooling systems allow in such cases the exploitation of thermal energy and the district heating infrastructure, which is already available on site.
- For the district heating system managing utility, thermally driven cooling applications allow enhancement of the amount of sold thermal energy and to reduce the seasonal differences in thermal energy demand.
- There is no need to implement a specific district tube system for chilled water, as the chilled water is produced on site.

In general, the thermally driven cooling technology is a small niche market in comparison to the overall air conditioning and cooling market. But attention is increasing, and specific market applications are being developed.

In Figure 8.16 the number of installed solar cooling applications worldwide is shown as one example of the application of thermally driven chillers. As can be seen most installations are still found in Europe. This is a sign that the technology and market are still in a very early stage of development. By becoming more mature, it is expected that most systems will be installed in the large solar thermal markets (e.g., China) and in markets with large cooling and air conditioning applications.

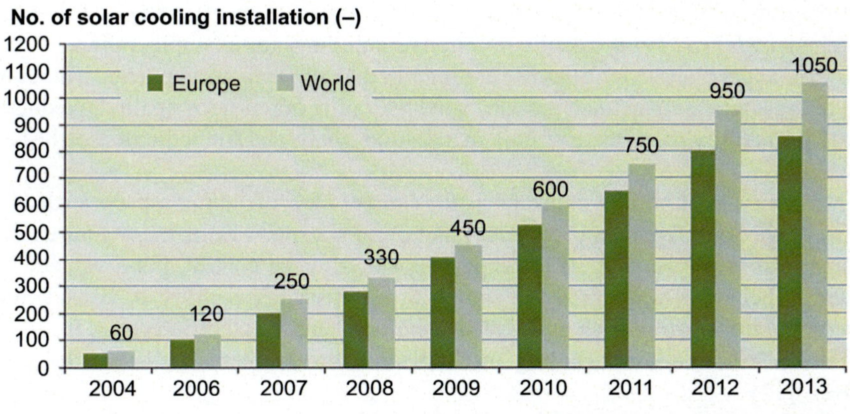

Figure 8.16 Market development of solar thermal air conditioning and cooling applications from 2004 to 2013.
Source: Climasol, EURAC, Fraunhofer ISE, Green Chiller, Roccoco, Solem Consulting, Tecsol [4].

8.3 Research and development needs and future trends in technological development and applications

Because hybrid systems combine more than one energy source and energy technology into a single system, they always profit from the improvements of single-energy technologies. Regarding the research and development needs for single technologies, the reader is invited to refer to the relevant chapters in the present book and the given links.

The **efficiency of hybrid systems** depends on the proper integration of the single technologies into one system. Every first-of-a-kind integration of technologies carries the risk that the efficiency remains below expectations and theoretical upfront calculations. To allow the proper functioning of hybrid systems substantial experience with each system combination is needed; this might include dynamic system simulations, first installation of systems, monitoring of consumption data, data evaluation, systems improvements, further improved installations, further monitoring, and so on.

To allow such a development in **small-scale hybrid systems**, two specific aspects are regarded by the authors as highly relevant:

- ***Execution of dynamic tests of the full system***, to include in the test not only the single components but also the hydraulics of the system, the settings of the control algorithm, and the management of the heat storage. Such tests are commonly carried out by single research laboratories, but a European or international standard test cycle was still missing when the present chapter was written.
- ***High degree of prefabrication*** and development of ***fully integrated kit systems***. Here through the hydraulics, the control unit and the dimensioning of all single components can be done and tested by the system developer. The probability of mistakes in installation and the installation time can be dramatically reduced.

Within **large-scale hybrid systems** and in system refurbishment toward hybrid systems such a prefabrication is not possible. To allow, nevertheless, to standardize the integration of different technologies and to profit from experiences from other systems, ***prefabricated hydraulic junctions***, including, for example, pumps, valves, heat exchangers, and control units, are regarded as relevant components. The objective of the application of such prefabricated hydraulic junctions is again the reduction of the probability of mistakes in the installation phase and the reduction of installation time.

A further aspect of high relevance is the ***control unit***, including aspects such as the control logic, measurement points, set points, and actuators. Here, many future developments are possible, such as dynamic adaptation of the control, improvement with customer behavior evaluation, integration in smart electrical grids, or consideration of external information such as weather conditions and weather forecast.

Further information regarding research and development needs can be found in Ref. [12].

Sources of further information

Important work in the field of renewable heating and cooling has been done within the many international multi-annual projects within the International Energy Agency

(IEA). For the present topic especially, the Solar Heating and Cooling and the Heat Pump Program are relevant.

Further information, technical reports and books can be found under:

- www.iea-shc.org/
- www.heatpumpcentre.org/

Websites from organizations and associations that might be helpful for the reader:

- www.rhc-platform.org/
- www.estif.org, www.egec.org, www.aebiom.org
- www.euroheat.org, www.ehpa.org, www.greenchiller.eu

Regarding technical handbooks, the following books summarize the main outcomes of two IEA projects:

- *Solar cooling handbook—A guide to solar assisted cooling and dehumidification processes*, edited by H. M. Henning, D. Mugnier, M. Motta, ISBN: 978-3709108413
- *Solar and heat pump systems for residential buildings*, edited by Jean-Christophe Hadorn, ISBN: 978-3-433-03040-0

References

[1] Low energy building in Branzolo, Province of Bolzano, Northern Italy. Building set up by IPES; demo heating and cooling installation set up under the coordination of EURAC within the European FP7 project ALONE. http://www.eurac.edu/en/research/technologies/renewableenergy/newsandmedia/Pages/photogallery.aspx?entryid=108687.

[2] Frequently asked questions about district heating and cooling, Euro Heat & Power, http://www.euroheat.org/District-Energy-Explained-210.aspx.

[3] District heating system of Chiusa, Province of Bolzano, Northern Italy, SEL Group, http://www.sel.bz.it/en/unternehmen/news-media/fotoarchiv.html.

[4] International Energy Agency – IEA, Solar Heat Worldwide 2015, http://www.iea-shc.org/solar-heat-worldwide.

[5] European Heat Pump Association, Summary of market figures 2014, http://www.ehpa.org/market-data/2013/.

[6] W. Sparber, K. Vajen, S. Herkel, J. Ruschenburg, A. Thür, R. Fedrizzi, M. D'Antoni, Overview on solar thermal plus heat pump systems and review of monitoring results, ISES Solar World Congress, 2011.

[7] International Energy Agency – IEA: Solar Heating and Cooling Program Task 44, Heat Pump Program Annex 38, http://task44.iea-shc.org/.

[8] J.-C. Hadorn, Solar and heat pump systems for residential buildings, 2015, ISBN 978-3-433-03040-0. http://eu.wiley.com/WileyCDA/WileyTitle/productCd-3433030405.html.

[9] International Energy Agency, SHC Task 44, HPP Annex 38, Technical report, A report of market available solar and heat pump systems, in: J. Ruschenburg, S. Herkel (Eds.), http://task44.iea-shc.org/publications.

[10] District heating and cooling country by country survey, Euro Heat & Power, 2015. http://www.euroheat.org/Publications-8.aspx.

[11] European Geothermal Energy Council EGEC, Market report 2013/2014 update, http://egec.info/media/.

[12] Strategic research priorities for cross cutting technologies, European Technology Platform Renewable Heating and Cooling, http://www.rhc-platform.org/fileadmin/Publications/Crosscutting_Strategic_Research_Priorities.pdf.

Renewable district heating and cooling technologies with and without seasonal storage

9

Jan Erik Nielsen, Per Alex Sørensen
PlanEnergi, Denmark

9.1 Introduction

District heating/cooling has several advantages:

- One or a few large central production facilities are much cheaper than thousands of individual installations.
- Some heat resources can only be utilized economically in district heating/cooling—e.g., excess heat from industrial production, waste heat from electricity production, heat from waste incineration, geothermal energy.
- District heating and cooling is the only way to supply (existing/historical) city centers with large fractions of renewable heating/cooling (renovation to zero-energy houses is not possible here).

Furthermore, it is not easy to integrate many small individual systems in an aesthetic way in cities—see Figures 9.1 and 9.2.

The use of renewable energy technologies is developing very quickly in the district heating sector as seen in Figure 9.3. One of the reasons for this rather fast transition is

Figure 9.1 Individual air-conditioning systems in Singapore.
Photo: Jennifer Steele, www.airconco.com.

Renewable Heating and Cooling. http://dx.doi.org/10.1016/B978-1-78242-213-6.00009-6

Figure 9.2 Solar water heaters supplying individual apartments in a city in China. Source: www.greenlawchina.org.

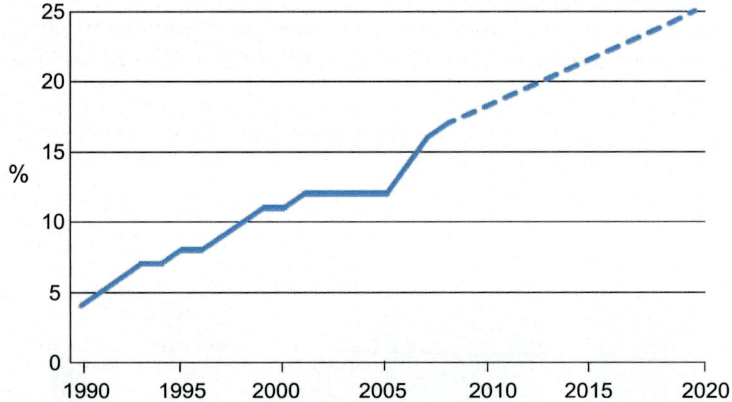

Figure 9.3 Share of renewable energy in district heating and cooling in Europe. From Ref. [1]. Data provided by Euroheat & Power.

the fact that district heating in general is very flexible in the use of different fuels and in the change from one fuel to another. Instead of exchanging thousands of small boilers in the individual building, exchange/modification can be done in one (or a few) central places. Therefore, when the right economic and political conditions are in place for a change, this change can happen very rapidly in the district heating sector. Another reason is that district heating and cooling (DHC) are very comfortable for the customer and need only a minimum of maintenance and involvement from the user side.

In Figure 9.4 the use of different energy resources in district heating is illustrated. Solar thermal, geothermal, and biomass are normally seen as the renewable energy

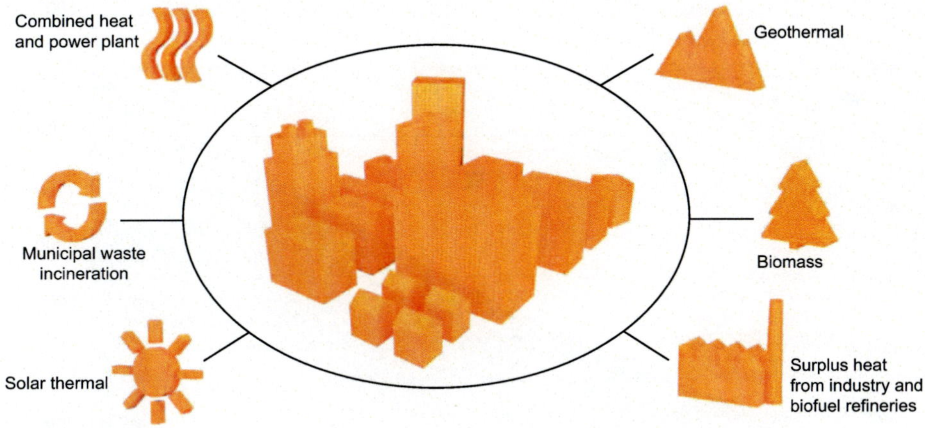

Figure 9.4 District heating sources.
From Ref. [4].

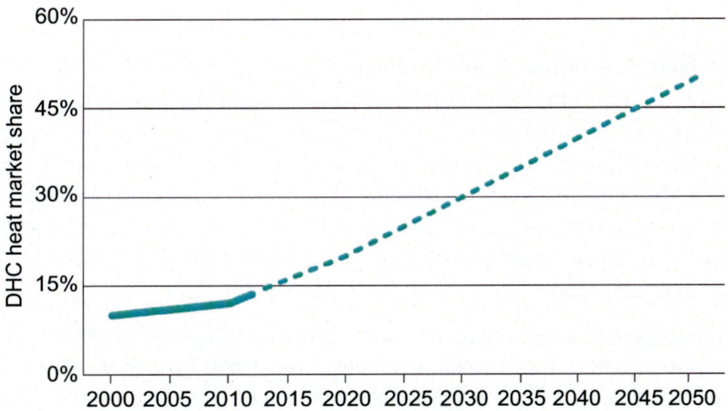

Figure 9.5 The market share of district heating is expected to grow.
From Ref. [1].

sources, but some (increasing) parts of the waste heat from electricity power plants and industry—and heat from waste incineration origins from renewable energy sources as well. Heat pumps/chillers too, more and more use renewable electricity or renewable heat.

Now, the district heating market share is rather low in Europe: around 15% of the total heating and cooling demand. However, it is expected that this share will grow in the coming years. As seen in Figure 9.5, district heating might in the future cover more than half of the heating and cooling market.

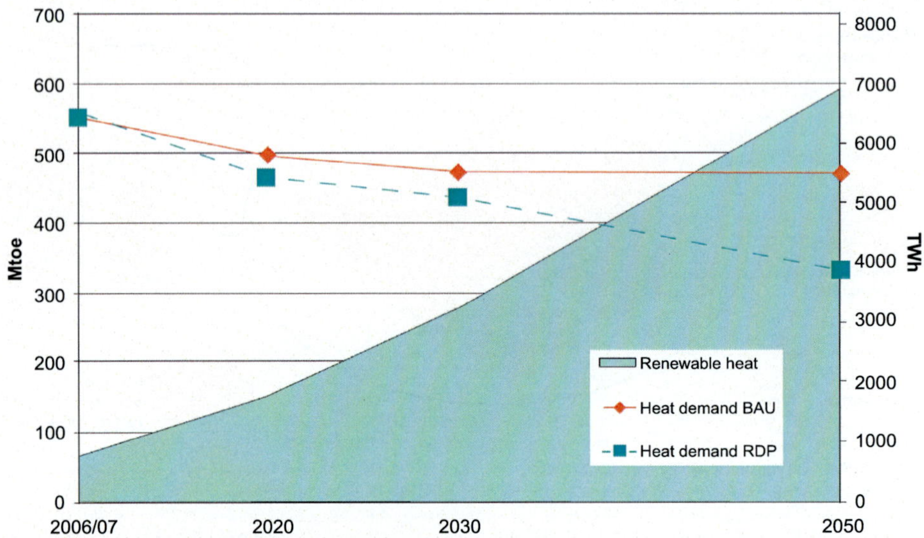

Figure 9.6 Heating supply from renewable energy sources in the EU. (From Ref. [3].) It is seen that more than enough renewable heat can be available around 2040. BAU is "Business as usual" scenario for heat demand; RDP is "Full research, development and policy" scenario for heat demand.

Another tendency influencing the development of renewable district heating is the decreasing demand for heating in general due to improved energy performance of the buildings—and extends to some industry processes too. In Figure 9.6 the decreasing heat demand is illustrated with the orange and blue lines.

Looking at the three tendencies together—increasing share of renewables in district heating, increasing market share of district heating, and decreasing heat market—the result expected is much more use of renewable district heating. Estimates based on Figures 9.4–9.6 are:

- 2020: Approximately 3 times more renewable district heating than in 2010
- 2030: Approximately 7.5 times more renewable district heating than in 2010
- 2050: Approximately 15–20 times more renewable district heating than in 2010

This dramatic increase in renewable energy will be covered by a variety of different sources. In Figure 9.7 the expected mix in the *total* heating market is shown.

At present, the cooling demand is below 5% of the total heating and cooling demand—this is expected to increase to around 20% in 2050. The share of district cooling might be higher, as DHC networks typically cover densely populated areas with offices, hospitals, etc., having a significant cooling demand. In Ref. [2] it is estimated that district cooling can supply 10% of the cooling demand in 2030 and 20% in 2050. Thus, the market potential for renewable district cooling is considerably lower than that for renewable district heating.

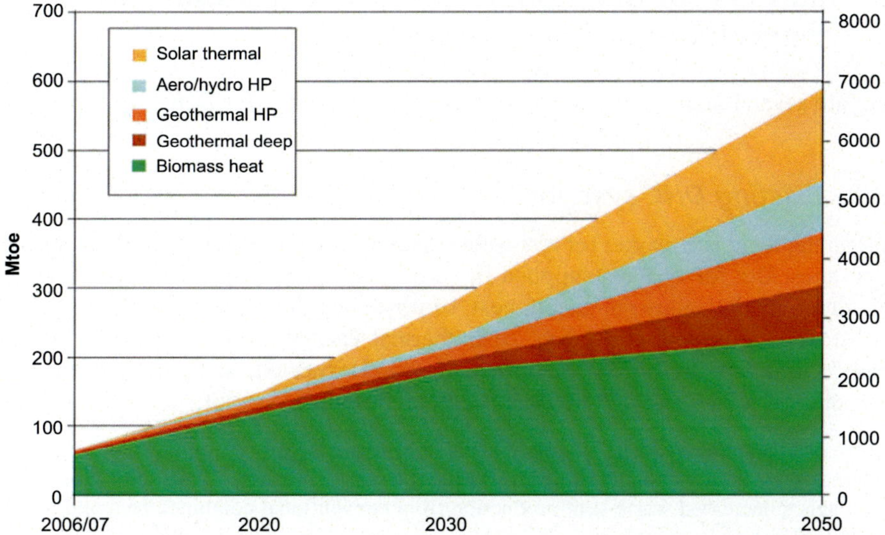

Figure 9.7 Shares of the different renewable heating sources in the EU. (From Ref. [3].) From 2040, when the demand could be covered (see Figure 9.6), a reduction in use in especially the biomass heat could be expected.

- Potential for renewable district heating 2050: 50% of total heating and cooling market
- Potential for renewable district cooling 2050: 4% of total heating and cooling market

Reference [3] gives the following summarized outlook for an expanding European DHC sector utilizing waste heat and renewable energy sources:

2020
- Avoidance of 9.3% of all European CO_2 emissions by district heating
- Additional 40–50 million tons of annual CO_2 reductions by district cooling
- Decrease of primary energy consumption with 2.14 EJ (595 TWh) per year, corresponding to 2.6% of entire European primary energy demand
- 25% share of renewable energy in district heating
- Reduced European energy import dependency with 4.45 EJ (1236 TWh)

2030
- A smart energy exchange network, allowing for optimal resource allocation between the multiple low carbon energy sources feeding into the system and various temperature demands of customers

2050
- Fully carbon-neutral energy solutions through regional, integrated networks

9.2 Markets for and applications of renewable heat generation for DHC—in general

There are several applications for renewable heat generation for DHC.

9.2.1 Existing DHC systems

In principle, there is a market for renewable district heat generation in all district heating systems. But for systems supplied by "waste heat"—i.e., heat from either waste incineration plants, industrial excess heat, or waste heat from power plants, it can be difficult to introduce, for instance, solar thermal energy because the base load is already "occupied" by cheap heat. However, it is expected, in the long term, that the amount of waste heat will be dramatically reduced because of:

- Increased recycling of waste → less waste will be available for heat production.
- Increased electricity production from wind and solar → decreased electricity production and hence decreased waste heat production from conventional combined heat and power (CHP) plants.

9.2.2 Areas now supplied by natural gas

Cities/suburbs/towns/apartment blocks now supplied with a natural gas grid, can be supplied by renewable district heating, as such areas most often have a building density justifying a district heating network.

9.2.3 City centers

For city areas that cannot be renovated to a low- or zero-energy standard (e.g., historical city centers), DHC is the only way to supply the necessary heating and cooling—this supply should be based on renewable energy or waste heat.

9.3 Resources for renewable DHC

Looking from the perspective of the different resources, there are only a few limitations for use of renewable district heating.

9.3.1 Biomass resources

Biomass can always replace fossil fuels used in plants producing DHC (including district heating from CHP plants). Whether this is then actually done is mainly a matter of

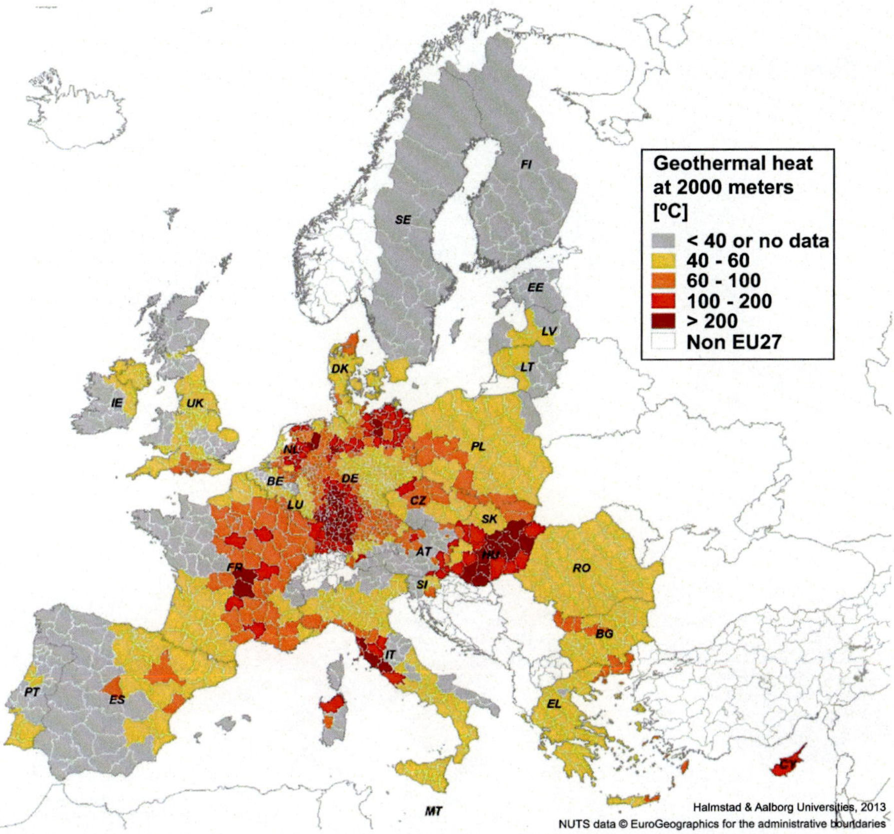

Figure 9.8 Identified geothermal heat resources by temperature at 2000 m depth by Nomenclature des unités territoriales statistiques (NUTS)3 region. From Ref. [2].
Source: European Commission, Atlas of Geothermal Resources in Europe. Publication EUR 17811, Luxembourg 2002.

fuel prices. The biomass resources are relatively widely spread in Europe (except from the very South)—and biomass can be transported.

9.3.2 Geothermal resources

The market for deep **geothermal heat** is normally limited to district heating systems close to areas with high temperatures in the underground. The larger the heat production the longer distance the heat can be transported. Figure 9.8 shows the areas where geothermal energy is/will be available.

9.3.3 Solar resources

The commercial market for **solar district heating** is so far mainly limited to smaller towns (with 1000–10,000 inhabitants) in the countryside with land available and an

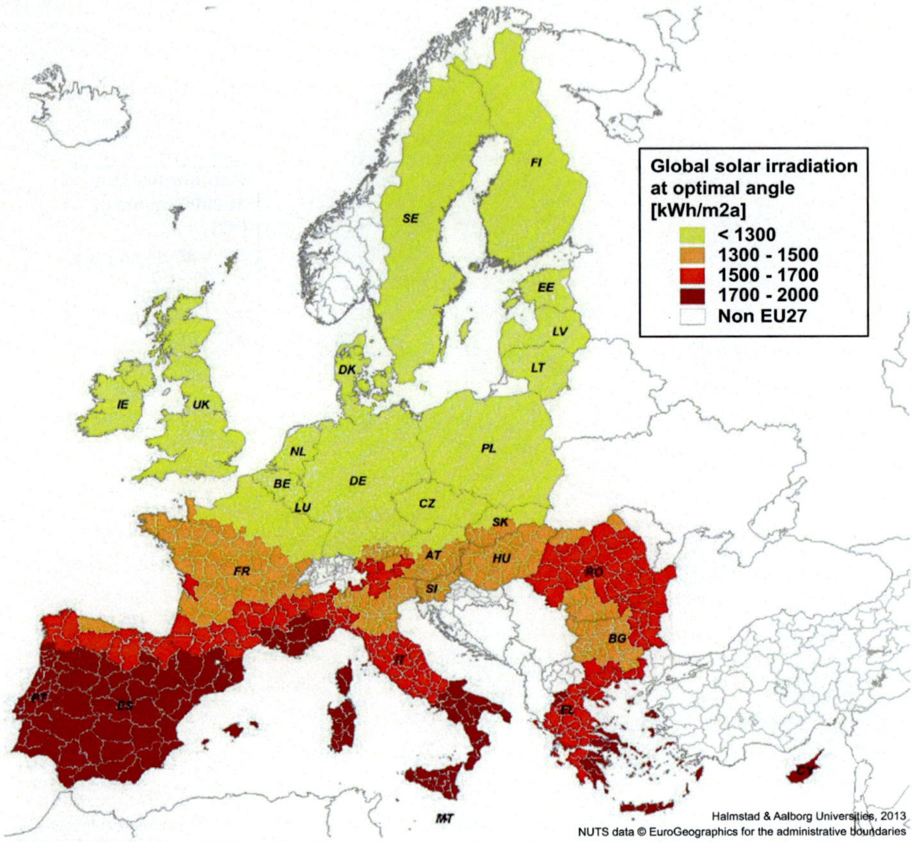

Figure 9.9 Annual solar irradiation on a south-oriented tilted surface at optimal angle. From Ref. [2].

already existing district heating network. In the longer term, major cities could also be supplied by solar district heating by:

- Very large ground-mounted collector fields (e.g., 100,000–1,000,000 m^2 or 70–700 MW) delivering heat from relatively long distances, e.g., connected to transmission pipes between cities.
- Roof-integrated collector fields in cities and industrial areas near cities delivering to the district heating network.

Without heat storage, solar district heating can supply 10–30% of the annual demand depending on the annual distribution of the solar radiation and the load. With seasonal storage, solar district heating can supply up to 100% of the load. The distribution of solar resources is seen in Figure 9.9, but a very high solar radiation level is not necessarily required, as proven by the success of solar district heating in Denmark—Figure 9.10.

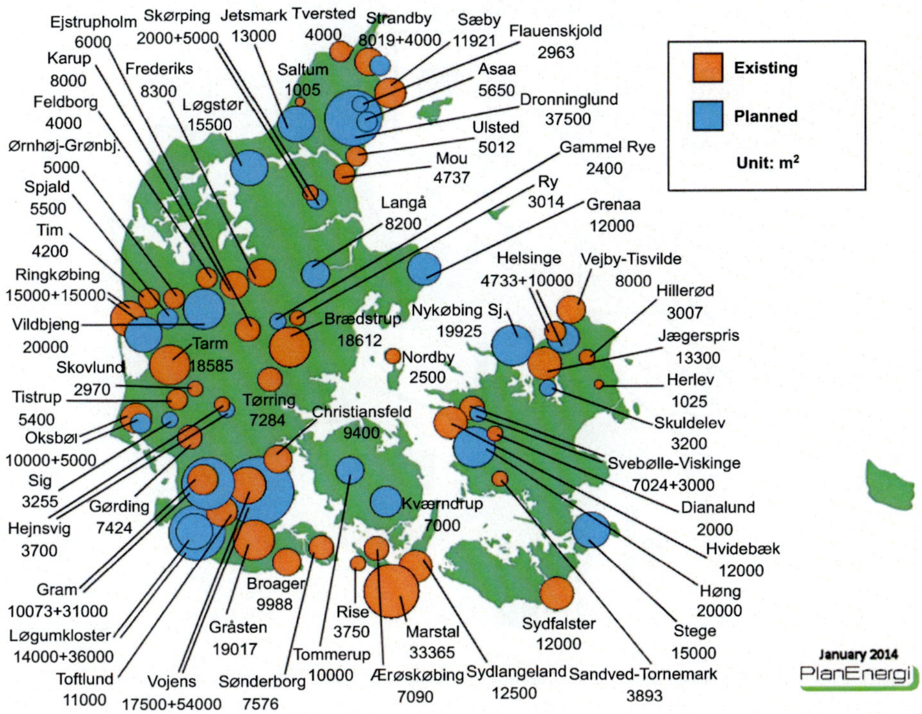

Figure 9.10 Solar district heating systems in Denmark.
Source: PlanEnergi.

9.4 Heat/cold distribution technologies and systems

In **district heating systems**, the heat for warming buildings is produced in centralized heating plants and distributed in insulated pipes to the customers. The district heating network consists of transmission pipes between production facilities and cities and between cities, pipes in the streets, and connection pipes to the buildings. To heat buildings and for production of hot water, the supply temperature normally needs to be 60 °C.[1] In the buildings are substations transferring the heat from the district heating network to the local heat distribution system. Substations for local areas can be introduced as well. If the substations also have heat exchangers, because of heat losses in the pipes, the forward temperature from the district heating production plant has to be higher. In traditional European systems, the forward temperature from the heating plant very often is 90 °C, and the return temperature is 60 °C. In a typical Danish

[1] The 60 °C is needed for the hot water production—for space heating it is possible to go to much lower temperatures depending on the heat distribution system and heat load of the building.

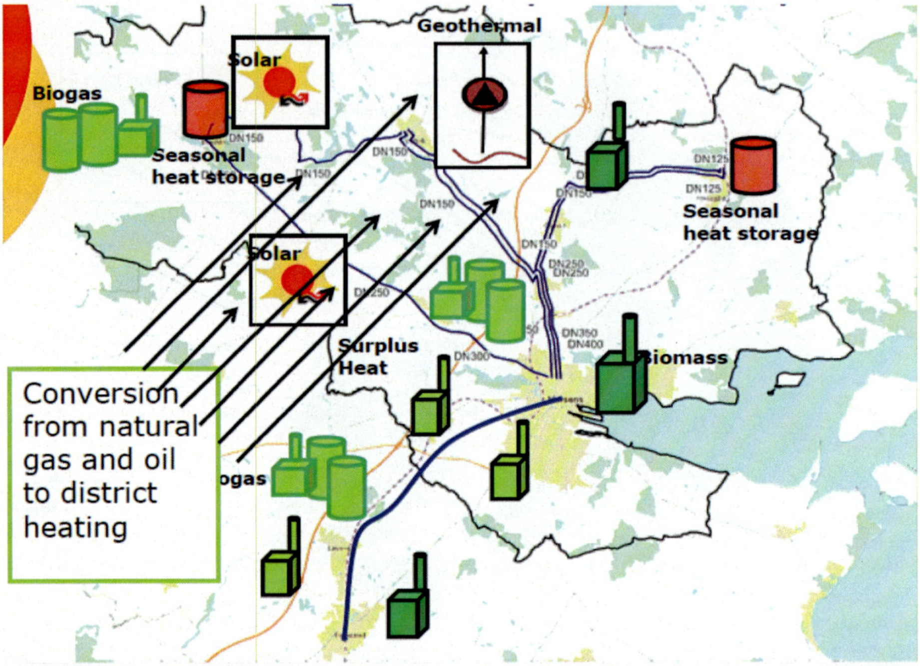

Figure 9.11 Flex Cities project in Horsens and Hedensted municipalities in Denmark. Large transmission pipes (up to DN400) connect the cities, towns, and the heat production facilities: excess heat, solar thermal plants, large storages, heat pumps, and biomass production plants. Source: Kristensen Consult Aps.

district heating system, the forward temperature normally is 75 °C, and the return temperature is 35–40 °C. In efficient district heating systems, the temperatures have to be as low as possible and the pipes well insulated. During the past 20 years heat losses from district heating pipes have been reduced by at least 50% because of better insulation and introduction of twin pipes, in which forward and return pipes are within the same insulation. This, combined with reduction of the temperatures in the district heating network, makes modern district heating systems very efficient and allows also longer transmission pipes to gain waste heat, excess heat from industries, and central solar thermal plants. An example of that is the Flex Cities project in Horsens and Hedensted municipalities in Denmark, where transmission pipes connect the cities, and excess heat, solar thermal plants, large storages, heat pumps, and biomass production plants are feeding in from different locations in the network (Figure 9.11).

In **district cooling systems**, cold water is produced at centralized production facilities (free cooling from seawater or groundwater), electrically driven cooling (compressor cooling) or heat-driven cooling (absorption cooling). The cold water is distributed in pipes in the streets. The users are shopping centers, office buildings, hotels, and other buildings, for which cooling is needed. District cooling networks can, for instance, be found in Göteborg and Stockholm in Sweden and in Helsinki in Finland.

9.5 Renewable heating production—description of technologies

9.5.1 Solar thermal

There are two main system types for solar district heating:

- Large central ground-placed solar collector field
- Distributed/decentralized collector fields placed on buildings in the district heating network

See Figures 9.12 and 9.13.

The solar collectors used in the central large ground-placed solar district heating collector fields in Denmark are typically large flat-plate collector modules of 12 to 14 m^2 each (Figure 9.14). The collectors are typically mounted upon concrete blocks set directly on the ground (Figure 9.15). The modules are connected in series of 10 to 25 modules in a row, and then several rows are connected in parallel. The price for a collector field including piping, heat exchanger, and control "ready to operate" is around 200 €/m^2.

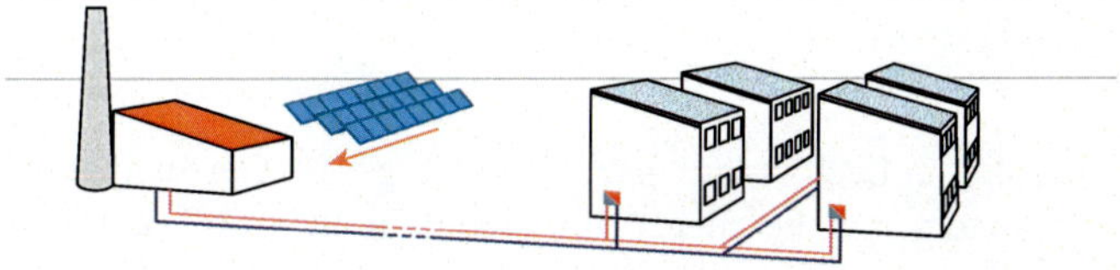

Figure 9.12 System with central collector field supplying centrally the whole network.

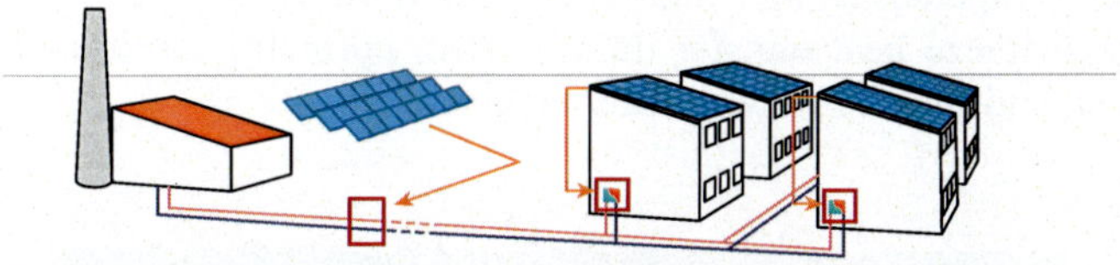

Figure 9.13 System with decentralized collector fields supplying substations in the network.

Figure 9.14 Large flat-plate collector module.
Source: ARCON-SUNMARK.

Figure 9.15 Field of ground-mounted solar collectors at Ulsted District Heating Company. Source: ARCON-SUNMARK.

Evacuated tubular collectors are used, too—e.g., in Germany and China—see Figure 9.16. Collectors can be nicely integrated in the roof of the building—see Figure 9.17.

The piping used for connecting the collector modules in the field shall be able to deal with high temperatures and high temperature variations. As there is a risk of freezing, an anti-freeze heat transfer fluid is used, normally based on 1.2 propylene glycol and with additives to prevent internal corrosion.

Figure 9.16 Large evacuated-tubular collector module for mounting on a flat roof. Source: Ritter XL.

Figure 9.17 Roof-integrated collectors.
Source: Austria Solar/ESTIF.

9.5.2 Biomass

Nearly all kinds of biomass can be utilized for combustion. Wood chips, wood pellets, dry waste wood, straw, and rice husk are typical examples. Biomass can be utilized in boilers with heat production only or in systems with CHP production.

9.5.2.1 Boilers with heat production

Large boilers for combusting biomass have efficiencies between 85 and more than 100% of the lower calorific value of the fuel. Efficiencies of more than 100% can be reached by condensing flue gas from the combustion process. Especially if wet fuels (up to 55% moisture of total weight) are used, the efficiency can be high. Boiler systems for heat production only can be combined with heat-driven heat pumps. The heat source for the heat pump can be flue gas (internal heat source) or external heat sources as excess waste heat (see subsequent discussion about heat sources for heat pumps).

9.5.2.2 Systems with CHP production

Systems with CHP production can be

- Boiler combined with an Organic Rankine Cycle (ORC)
- Boiler combined with steam turbine
- Gasification unit combined with gas turbine or gas engine

In **ORC systems**, the biomass boiler produces hot oil (about 310 °C) to heat up the ORC. The ORC is built as a steam turbine but using organic oil with a low boiling temperature instead of water. ORC systems are the most feasible CHP systems, with biomass for electricity outputs below 2 MW. Approximately 250 systems are implemented, most in southern Europe and Germany (Figure 9.18).

In **systems with steam turbines**, the biomass boiler produces pressurized hot water at approximately 500 °C for a steam turbine. Efficiencies can be up to 50% for very large systems using combined cycle technology. Biomass systems with steam turbines can be found in sizes from 1 MW electricity output all over the world.

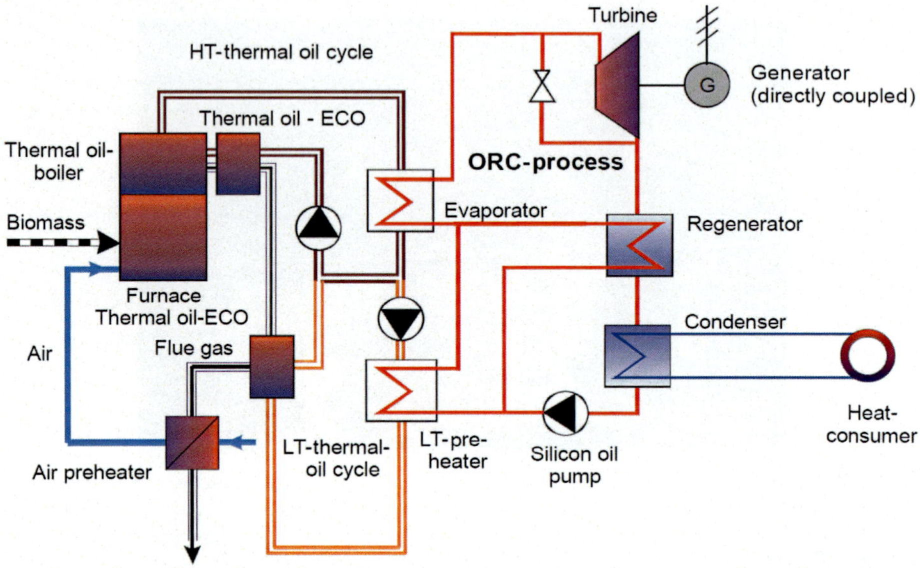

Figure 9.18 Main components of the biomass-fueled thermal oil boiler system including Organic Rankine Cycle.
Source: BIOS Bioenergiesysteme GmbH.

In **gasification** systems, the biomass is heated up without burning. The gasified gas is cleaned and can be used in a gas turbine or a gas engine. Total electrical output for large plants can be as high as the best steam turbine plants, and for smaller plants it can be more than 30%, but the technology is still under development, and reliable systems are difficult to find at the market.

9.5.3 Geothermal

Geothermal heat can be found in deep reservoirs many places in Europe (see Figure 9.8). The temperature of the reservoirs differs, for instance, in 2000 m depth from 40 to 50 to more than 200 °C. For utilization, two drillings are made. A drilling for pumping up the hot salty water and another drilling for injection of the water after utilization. For high temperatures, the water can be used in an ORC for electrical production and for district heating. For utilization by district heating, it is normally economically feasible to take as much heat as possible out of the water before reinjection using thermally driven or electrically driven heat pumps. Then the capacity of the drilling will be maximized and the number of drilling couples can be minimized.

Before drilling, careful investigations have to be carried out to estimate the chance of finding water in a porous layer and at sufficient temperature because drilling is very often expensive.

9.5.3.1 Heat pumps

Heat pumps will be increasingly used in district heating systems because they are needed as regulation of fluctuating electricity production ("power to heat") and because cold water is needed for district cooling. There are two main types of heat

Figure 9.19 1.5 MW_{heat} high-pressure screw ammonia heat pump in Braedstrup district heating system.
Source: Braedstrup district heating, PlanEnergi.

pumps: thermally driven and mechanically driven heat pumps. The heat pumps must be able to produce hot water at forward temperatures (70–90 °C) for the district heating system. Heat pumps using CO_2 or ammonium as refrigerant and able to produce at these high temperatures are available at the market (Figure 9.19).

The **heat-driven heat pumps** (normally absorption heat pumps) can be used in heat production systems with combustion processes in boilers or engines. Hot water at 80 to 160 °C is used as the heat source, and heat can be taken from the same sources as utilized by mechanically driven heat pumps (see following).

Mechanically driven heat pumps use electricity-driven compressors or compressors directly driven by a gas engine.

Heat sources/cooling sources for heat pumps for district heating can be

- Excess heat from industries
- Cleaned waste water from sewage plants
- Groundwater
- Surface water from lakes, rivers, and sea
- Geothermal heat
- Air
- Flue gas from combustion processes
- Solar heat in pit heat storages or borehole storages

9.6 Thermal storage technologies in DHC systems

Using thermal storage in DHC systems has several functions and advantages:

- Leveling out mismatch between heat production and heat load (e.g., in connection with solar heat production and CHP production)
- Optimizing system performance (e.g., in connection with biomass CHP plants)

Figure 9.20 5500 m^3 steel tank for short-term buffering, before insulation.
Source: Braedstrup district heating, PlanEnergi.

- Combination of several different heat technologies
- Interaction with electrical grid—leveling out also mismatch between renewable electricity production and load (connecting heat pumps and combined heat and power plants units to a thermal store "Power to heat")

9.6.1 Short-term storage

Short-term storage means storing heat for hours or days. This can level out hourly, diurnal, and weekly differences in production and load and is widely used in connection with CHP plants.

Short-term storages are normally insulated and weather-protected steel tanks like those seen in Figure 9.20. Putting the tank in vertical position increases the ability to create and maintain temperature stratification and hence a high usable temperature in the top of the tank. Most tanks of more than 500 m^3 are pressureless and protected against corrosion by a layer of nitrogen between the water surface and the tank top. Temperatures in unpressurized steel tanks are up to 95 °C.

9.6.2 Long-term storage (seasonal storage)

Long-term storage means storing heat for months and even from summer to winter. This can level out also seasonal variation in production and load and is slowly becoming more and more popular especially in connection with solar district heating.

But a large thermal store can do more than just store solar heat from summer to winter. Introducing a large-capacity thermal store in the district heating system gives extremely good conditions for a very flexible production from combined production units. To illustrate this, look at Figure 9.21.

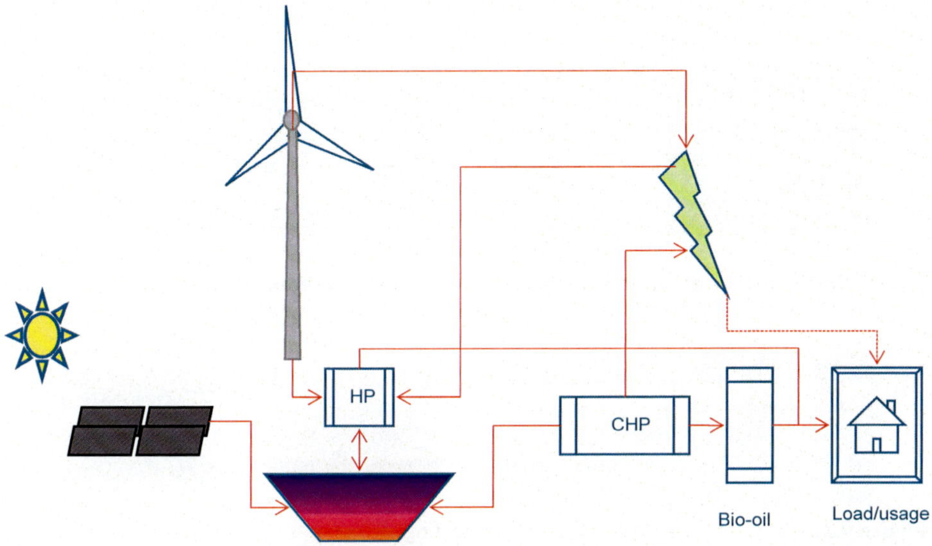

Figure 9.21 Schematics of a district heating system combining renewable heating technologies and heat storage and interacting with the electrical grid.
Source: PlanEnergi.

In the system schematically shown in Figure 9.21, the following heating technologies are combined:

- Solar collectors
- Heat pump
- CHP unit (e.g., gas motor)
- Boiler

Furthermore, the system is connected to the electrical grid (heat pump using electricity and CHP unit producing electricity).

Having high-capacity thermal storage, the benefits of the different technologies can be utilized.

Solar collectors

- Produce "free" solar energy—even when there is no heat load

Heat pump

- Produce cheap heat when electricity is inexpensive—even if there is no heat load
- Make fast capacity regulation (load)/earn money
- Reduce storage volume

CHP unit

- Produce valuable electricity/earn money
- Fast capacity regulation (production)/earn money

Four main types of long-term storage technologies are used (so far) in DHC systems—Figure 9.22.

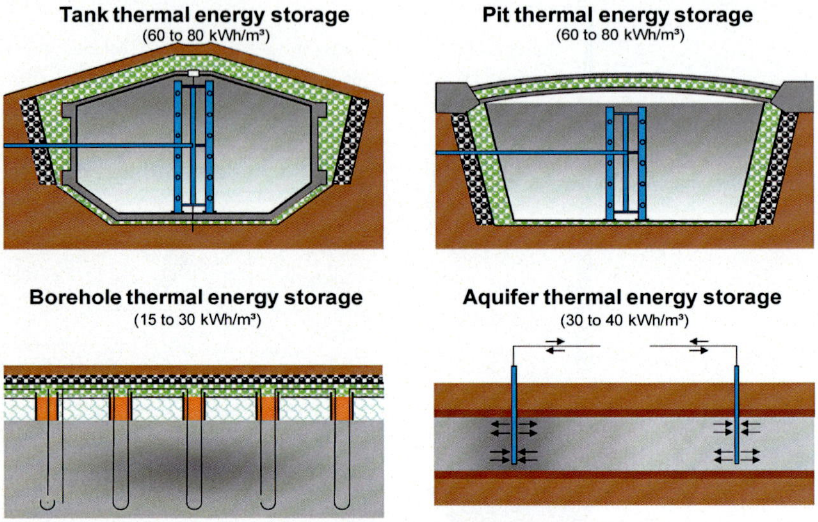

Figure 9.22 Four types of storage for district heating.
Source: SOLITES.

9.6.2.1 Tank thermal energy storage

Tank thermal energy storage (TTES) are often made from concrete and with a thin plate welded-steel liner inside. The type has primarily been implemented in Germany in solar district heating systems with 50% or more solar fraction. Storage sizes have been up to 12,000 m^3 (Figure 9.23).

Figure 9.23 Tank-type storage.
Source: SOLITES.

Figure 9.24 Pit thermal energy storage. New Marstal pit heat storage, 75,000 m^3.
Source: Marstal District Heating.

9.6.2.2 Pit thermal energy storage

Pit thermal energy storages are large water storages. The slope of the pit side is approximately 1:2, depending on the ground conditions. During excavation of the pit, excavated soil is used as banks and compressed. After excavation a plastic liner (HDPE) is implemented and welded to make the storage water tight, and finally water fills the storage and another plastic liner is welded around the margin, gradually pulled over the water and insulated (see Figures 9.24 and 9.25). The storages are not pressurized,

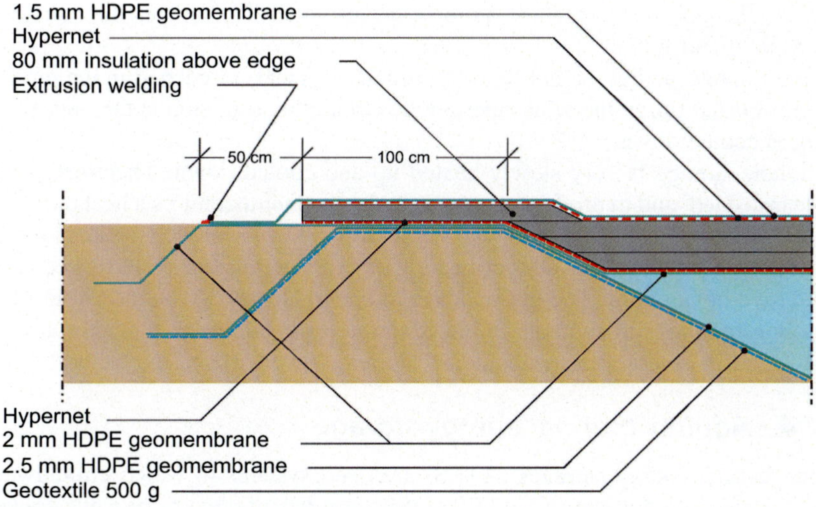

Figure 9.25 Details of pit heat storage construction.
Source: PlanEnergi.

Figure 9.26 Borehole thermal energy storage. Drilling boreholes.
Source: Braedstrup district heating, PlanEnergi.

and therefore the maximum temperature is 95 °C. Pit heat storages have temperature stratification as do steel tanks and tank storages.

Pit heat storages are built in a few places in Denmark. The largest is 75,000 m^3.

9.6.2.3 Borehole thermal energy storage

Borehole storage is a closed system, in which soil is heated up from hot liquid circulated in vertical pipes. A number of vertical boreholes are drilled. One or two U-pipes are placed in each borehole, and the borehole is filled with grouting that closes all gaps between pipes and the surrounding soil. The pipes are connected to a central connection well. On top of the borehole storage, an insulated lid finalizes the construction (Figures 9.26 and 9.27).

In the summer period the hot liquid (normally water) is circulated in the pipes, and the soil is heated up. In the winter period the circulation direction is the opposite, and the soil is cooled down.

Borehole storage is very slowly heated up and cooled down. Therefore, a buffer storage is needed, and to produce district heat forward temperatures a heat pump needs in most cases to be connected to the system. Only a few borehole storages for storing and producing hot water at district heating forward temperatures have been implemented (two in Germany, one in Canada, and one in Denmark), but borehole storages for cooling buildings in summer and heating the same buildings in winter have been implemented in several places, including Norway, Sweden, and the Netherlands.

9.6.2.4 Aquifer thermal energy storage

Aquifer thermal energy storages (ATES) are open systems in which groundwater is heated up and cooled down. An ATES system consists of couples of wells connected

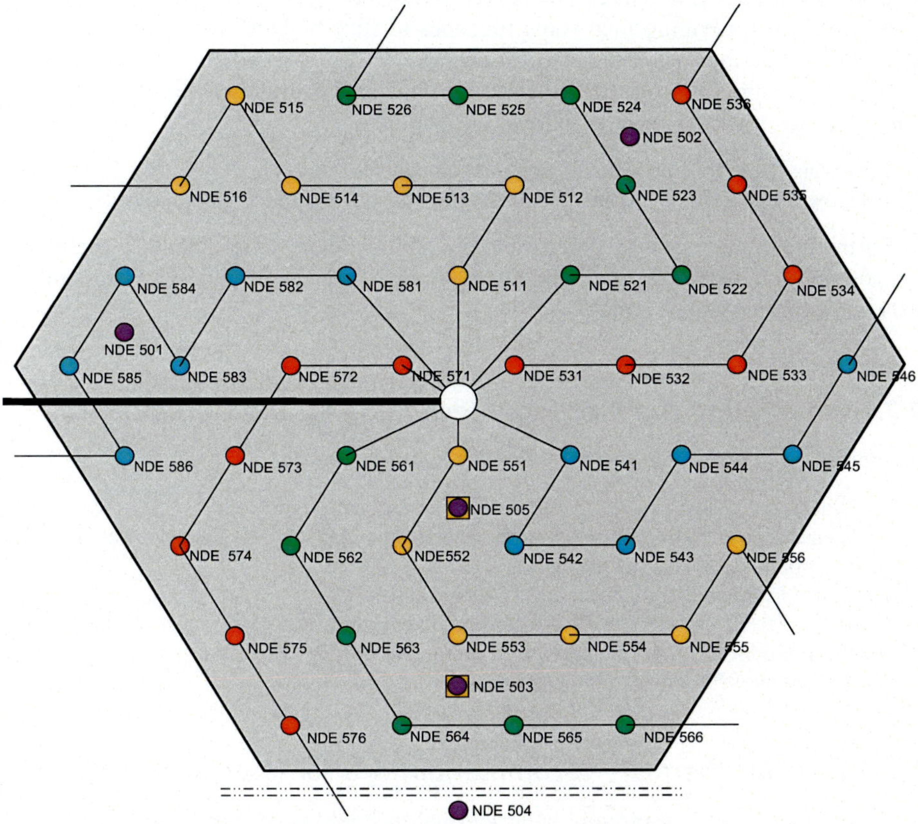

Figure 9.27 Borehole thermal energy storage. Layout of boreholes.
Source: Braedstrup district heating, PlanEnergi.

to the same groundwater reservoir. In winter the water in the reservoir is pumped up and cooled down in a heat exchanger, in which the other side is used to heat up buildings or as the heat source for a heat pump. The cold water is returned to the groundwater reservoir. In summer the direction is reversed and the cold water is heated up in a heat exchanger, in which the other side is used for cooling buildings or industrial processes.

ATES temperatures are from 2 to 30 °C. ATES is a proven technology used all over the world, but most are in the Netherlands, where more than 2000 systems are implemented.

9.7 Research and development needs and future trends in technological development and applications

Generally, the main challenge for renewable DHC is competitiveness with DHC generated by fossil fuels.

In principle, this problem can be solved politically by taxation of fossil fuels and CO_2 emissions (providing then some revenues to state budgets always in need—but at the same time increasing the price of energy).

Anyway, improving cost effectiveness of renewable district heating shall have priority, and two tracks should be followed in parallel:

- Improving the individual technologies and components
- Utilizing benefits from combining technologies and systems in an optimized/smart way

Furthermore, the focus for all technologies should be on environmental sustainability, and for biomass the focus should be on production/use/transport of biomass materials and emissions from boilers.

9.7.1 Improving the individual technologies and components

In general, cost/performance improvements can be done on:

- Investment costs (less expensive materials, more rational production).
- Maintenance costs (reliable products and systems, low operating costs (manpower and electricity), optimized safety precautions.
- Efficiency (improved component and system efficiencies).
- Upscaling (size: large systems have lower relative costs and losses than small systems; numbers: products/components produced in big numbers are less expensive).
- Skills (training of installers and operators).

9.7.1.1 Solar thermal—recommendations for improvement

Recommendations for improving solar thermal technology for DHC[2]:

- Improve efficiency of large-module solar collectors (flat plate and evacuated tubes) for ground mounting (optimize absorber and glazing coatings, optimize absorber design with respect to thermal efficiency, pressure loss, heat capacity loss, and material use, make larger modules, etc.).
- Reduce price of large-module solar collectors for ground mounting (materials, coatings, design, larger modules, rational production of big numbers, etc.).
- Reduce installation price of building-integrated solar collector (standardized solar collector building modules, simple prefabricated substations, experienced/trained installers, etc.).
- Optimize large-scale solar collector arrays for efficient and reliable operation (uniform flow distribution, low pumping power, absorber design, collector field layout, piping, operating strategies, efficient safety precautions, etc.).
- Improve optimize costs/lifetime for pipes operating with large temperature variations (standards, etc.).

[2] Inspired by Ref. [6].

9.7.1.2 Biomass—recommendations for improvement

Recommendations for improving biomass technology for DHC[3]:

- Improve efficiency of large biomass boilers (fuel mix, improved control, condensing, higher temperature for electricity production, etc.)
- Reduce price of large biomass boilers (materials, rational production of big numbers, etc.)
- Optimize large biomass boilers for efficient and reliable long life operation (reduce deposits, corrosion, materials, cleaning, control, etc.)
- Reduce emissions of CO, NO_x, dust
- Reuse of residues (ash to fertilizer)
- Biomass CO_2 capture and storage/utilization
- Efficient/sustainable production/supply chains of biomass

9.7.1.3 Geothermal—recommendations for improvement

Recommendations for improving biomass technology for DHC[4]:

- Mapping of resources (overview, details, etc.)
- Cheaper drilling (improved/new drilling methods, etc.)
- Improve reliability (component/instrumentation resistance to high temperatures/pressure and corrosion/scaling, etc.)
- Improve efficiency and lower costs on large-scale high-temperature heat pumps and chillers
- Improve CHP and heat and cold
- Enhanced Geothermal Systems (enhanced deep-drilling stimulation techniques/strategies, fracture mapping, transit time, transport paths, investigation/estimation, control of induced seismicity, rock–water interaction, etc.)
- Sustainability (measuring and understanding consequences for environment, etc.)

9.7.1.4 Heat storage—recommendations for improvement

Materials/construction for temperatures $\approx$ 100 °C.
Reliability

- Improved/new materials for storage construction and lining that can withstand 90 °C for 30 years
- Test of lifetime of liners
- Test of lifetime and ventilation of lid constructions
- Reliable monitoring results for demonstration plants
- Demonstration of (larger) storages
- Cheaper pit heat stores, e.g., by use of existing infrastructure (gravel pits, dry docks from closed shipyards)
- Demonstrate storage technologies for multifunctionality and new purposes (surplus heat from industrial processes, incineration plants)
- Demonstration of cheap storages for low-temperature purposes (below 50 °C)
- Integration of heat pumps with high COP and high supply temperature
- Comprehensive planning and realization process

[3] Inspired by Ref. [5].
[4] Inspired by Ref. [7].

Source of further information

www.solar-district-heating.eu is a very informative website with respect to solar district heating.

References

[1] Strategic Research and Innovation Agenda for Renewable Heating & Cooling, European Technology Platform on Renewable Heating and Cooling, March 2013.
[2] Cheaper Comfort, Faster Decarbonization, Better Energy—Heat Roadmap Europe, Second Study for the EU27, By Aalborg University, Halmstad University, Ecofys Germany GmbH, PlanEnergi—Euroheat & Power, May 2013.
[3] Common Vision for the Renewable Heating and Cooling Sector in Europe: 2020—2030—2050, European Technology Platform on Renewable Heating and Cooling, 2011.
[4] Frequently Asked Questions about District Heating and District Cooling, Euroheat & Power, www.euroheat.org.
[5] Strategic Research Priorities for Biomass Technology, European Technology Platform on Renewable Heating and Cooling, 2012.
[6] Strategic Research Priorities for Solar Thermal Technology, European Technology Platform on Renewable Heating and Cooling, 2012.
[7] Strategic Research Priorities for Geothermal Technology, European Technology Platform on Renewable Heating and Cooling, 2012.

Solar thermal heating and cooling in China

10

Ruicheng Zheng [1], *Zhiqiang Yin* [2]
[1]Institute of Air Conditioning, China Academy of Building Research (CABR), Beijing, China;
[2]Tsinghua University, Beijing, China

10.1 Introduction

10.1.1 Historical review (Xu et al., 2009)

Solar thermal heating and cooling in China has developed since the 1950s and has undergone three stages: starting, industrializing, and spreading.

1. Starting stage: The earliest solar water heating systems of China were developed in 1958 and were used in the public bathrooms in Tianjin University and Beijing Tiantanghe Farm. But it was only a first step for development and developed slowly from 1960 to 1980.
2. Industrializing stage: During the 1980s, China's science research results directly promoted the formation of the solar thermal industry, especially the manufacturing technology of Al-N/Al-selective coating and all-glass evacuated solar collector tubes that were developed by Tsinghua University and by many manufacturers through technical transformation. So China completed the solar thermal industrialization in the middle of the 1990s and formed the largest production and market scale for solar thermal in the world until 2000.
3. Spreading stage: Since 2000, solar thermal has spread very fast via market requirements and government encouragement in China, not only in the quantity of manufacturers and application but also in the level of technical progress. Nowadays China has become the country having the largest production output and installation for solar collectors in the world.

10.1.2 Characteristics of solar thermal in China

Development for solar water heating in China has its own typical characteristics compared to other countries, especially developed countries. These characteristics are mainly reflected in three aspects: market, products, and systems.

1. Market: Different from developed countries, no hot water was supplied to residential buildings in China before 2000, and citizens had to acquire a water heater to provide a scouring bath. As no gas was supplied to residential buildings of small cities or towns at that time, and the operational cost of electric water heaters was rather high, solar water heating systems were just right for satisfying the requirements of citizens in these cities and towns. Therefore, the development of solar water heating has been totally dependent on the market in China.
2. Products: Different from developed countries where flat-plate solar collectors take the main market share, evacuated-tube solar collectors take more than 90% market share in China.

Renewable Heating and Cooling. http://dx.doi.org/10.1016/B978-1-78242-213-6.00010-2

Figure 10.1 Typical close-coupled domestic solar water heating system in China.

In developed countries, remote-storage domestic solar water heating systems with forced circulation are mostly used, but in China close-coupled domestic solar water heating systems with natural circulation are more popular. Figure 10.1 shows the typical close-coupled domestic solar water heating system used in China. But this type of domestic solar water heating system will gradually decrease in the market of cities in the future as the requirement of building integration for solar water heating systems develops in cities.

3. Systems: In recent years the policy of compulsory installation for solar water heating systems was issued in many Chinese cities. But as most residential structures are tall or multistory buildings in the cities, the solar water heating systems integrated into these buildings should be a central system for an entire building's hot water supply or a remote-storage system for one apartment's hot water supply (in general, the solar collectors in this system are installed on the balconies). Figures 10.2 and 10.3 show these two typical systems.

Figure 10.2 Central solar water heating systems installed on the roofs of the tall buildings.

Figure 10.3 Remote-storage solar water heating system installed on balconies of apartments of a tall building.

Solar water heating systems take the largest market share and application ratio of solar thermal application in China, but in some European countries solar heating combisystems take more share and ratio. This is also a differing characteristic of China's solar thermal application.

10.2 Market situation and development

10.2.1 Solar thermal manufacturers

Nowadays in China there are over 2800 solar thermal manufacturers, which include the largest 4–5 leading enterprises, 20 large enterprises, and over 100 local backbone enterprises; for the distribution of these manufacturers, see Figure 10.4. Various leading enterprises have taken over 40% market share of China till 2012 and more than 30 leading enterprises have production value over 100 million RMB Yuan per year. The leading enterprises have strong capacity for research and development (R&D), modern facilities, and international competitiveness; their export value reached 300 million USD, and export areas are more than 200 countries in 2012. For the yearly production output for solar collectors in China from 2001 to 2013, see Figure 10.5. The yearly growth rate has been slower than in the past prior to 2011.

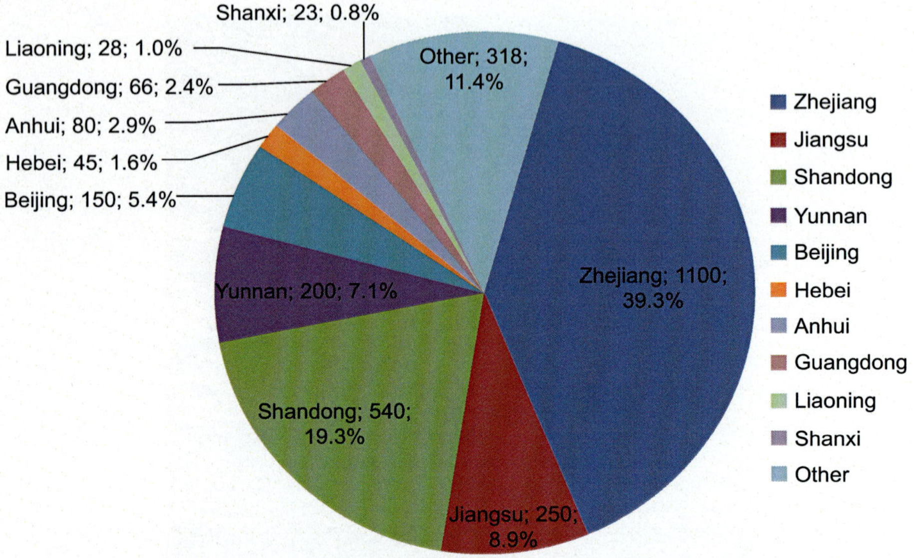

Figure 10.4 Distribution of manufacturers in China.

Solareast Corporation, which has the Sunrain & Siji Micoe trademark, Co. Ltd is the first listed company of solar thermal in China. Some other important solar thermal manufacturers are Himin Solar Energy Group Co. Ltd, Shandong Sunle Solar Energy Co. Ltd, Shandong Linuo-paradigma New Energy Resource Co. Ltd, Beijing Tsinghua Solar Ltd, Beijing Tianpu Solar Energy Industrial Co. Ltd, Jiangsu Huayang Solar Energy Co. Ltd, Jiangsu Sunxia Solar Energy Industrial Co. Ltd, and Guangdong Fivestar Solar Energy Co. Ltd.

10.2.2 Applications for solar thermal heating and cooling

In 2013 the installed capacity of solar thermal in China reached 310 million m^2 solar collectors, accounting for about 60% of the world's total; Figure 10.6 shows the yearly installation capacity for solar collectors from 2001 to 2013 in China.

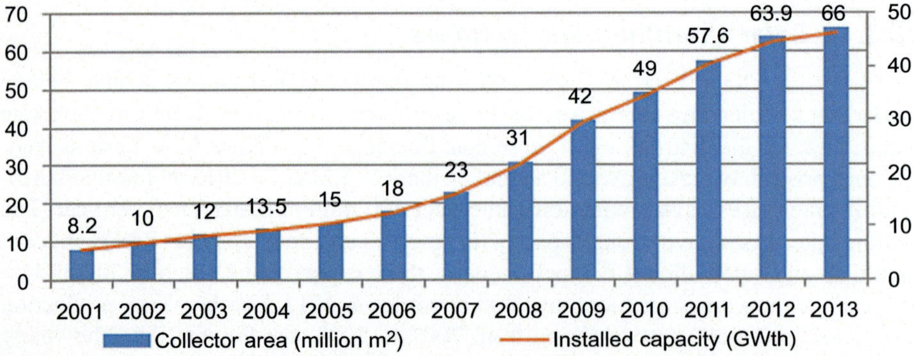

Figure 10.5 Yearly production output for solar collectors in China (2001–2013).

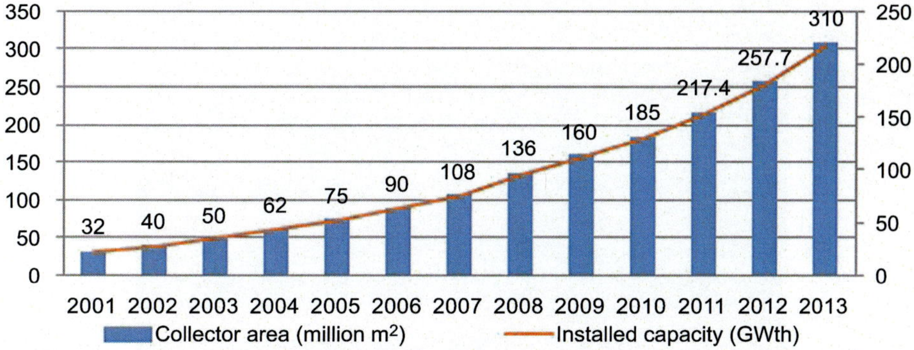

Figure 10.6 Yearly installation capacity for solar collectors (2001–2013).

In China, there are mainly three kinds of applications for solar thermal in buildings: solar hot water systems, solar heating combisystems, and solar cooling systems. Solar hot water takes the largest application share: about 92%, solar heating combisystem is about 7%, and solar cooling is about 1%.

1. Solar water heating system
There are two kinds of application markets for solar water heating systems in China. One is in cities where solar water heating systems integrated in buildings are needed. Another is in the countryside where close-coupled solar water heaters can be used. In the past, close-coupled solar water heaters took a larger market share in China. However, solar water heating systems integrated into buildings in cities will increase, and this type of system will be more popular in the future in China. As most buildings are high-rise structures in the cities of China, the suitable solar water heating systems integrated in these buildings are central systems installed on the roofs or remote-storage systems installed on the individual family balconies. However, in the towns and countryside of China most buildings are single-story or low-rise buildings, so the suitable solar water heating systems that are integrated in buildings of towns are central systems installed on the roofs, and close-coupled solar water heating systems can be still used in the countryside of China.
2. Solar heating combisystem
The situation is different for application of the solar heating combisystem in China and some European countries. In some European countries, especially Denmark and Sweden, solar heating combisystems are more abundant than solar water heating systems, but in China the application share for the solar heating combisystem was fewer in the past. Now the situation is changing but developing slowly. As China's special condition (more high-rise buildings and lack of land), there are two suitable application areas for solar heating combisystems nowadays in China. One important application is in the rural areas of eastern China where the peasants have higher economic capacity and standard of living such as the countryside in the suburbs of Beijing, where over 300,000 m^2 of building space using solar heating combisystems has been built in recent years. The other important areas are the western areas in China such as Gansu and Qinghai Provinces, Inner Mongolia, and Tibet Autonomous Regions, etc., where having better solar resources, cold weather, thinner population, larger uncultivated lands, and the Gobi Desert are suitable to build large solar heating combisystems with seasonal heat storage. Figure 10.7 shows the solar heating combisystem with seasonal heat storage in Urad Zhongqi of Bayannur City of the Inner Mongolia

Figure 10.7 Solar heating combisystem with seasonal heat storage in the Inner Mongolia Autonomous Region of China.

Autonomous Region. In western areas of China, the government should give some financial support to build solar heating combisystems. But in eastern areas solar heating combisystems can be built through the market and governmental promoting policy.

3. Solar cooling system
Compared with solar water heating and heating combisystems, solar cooling needs higher working temperatures and the initial investment cost, so its application share is now the lowest in China. Until the end of 2013 there have been only a few demonstration projects for solar cooling in China, and parts of them were built in buildings owned by solar enterprises. But a few solar cooling projects for public buildings show very nice demonstration effects, such as the solar air-conditioning system in the Rear-Service Building of the International Center for Competition of Sailing Boat in the 29th Olympic Games in Qingdao and the solar air-conditioning system in the library of Hainan University (see Figure 10.8); the solar fraction of these two projects can reach 25–30% in summer. If one is concerned with weather, different areas have different demands for solar cooling applications in China. Nowadays there are two suitable areas for solar cooling in China. One is in the hot summer and warm winter zone in which cooling is needed almost all year, so solar air conditioning is only for cooling in these areas. Another is in the hot summer and cold winter zone in which cooling is needed in summer and heating is needed in winter, so solar air conditioning is for cooling in summer and heating in winter in these areas.

The cost of investment for solar heating and cooling in China is lower than in many countries. The cost range of different systems now is as follows:

1. Close-coupled solar water heating system: 1000–2000 Chinese Yuan Renminbi (CNY)/m^2 collector
2. Remote-storage solar water heating system installed on the balcony of an apartment: 2500–3000 CNY/m^2 collector
3. Central solar water heating system installed on the roof of a building: 900–2500 CNY/m^2 collector
4. Solar heating combisystem with short-term heat storage: 2500–3000 CNY/m^2 collector
5. Solar heating combisystem with seasonal heat storage: 3000–4000 CNY/m^2 collector
6. Solar cooling system: 2500–3000 CNY/m^2 collector

Figure 10.8 Solar cooling system at Hainan University.

The main driver for development of solar thermal in China is the historic market requirement, but as serious pressure increases for environmental protection and the changing climate, the government has paid more attention to solar thermal in recent years, and some governmental policies have become another key driver of solar thermal development. For promoting energy savings and decreasing CO_2 emissions, China's central and local governments have issued some policies to support renewable energy application in China. The main policies are as follows:

1. Compulsory installation for solar water heating systems:
   Policies of compulsory installation for solar water heating (SWH) systems were issued in many cities and provinces of China from 2006. The demands for the SWH compulsory installation in some provinces such as Jiangsu, Shandong Provinces, etc., are solar water heating systems shall be installed in new buildings that are lower than 12 stories. But the demand for SWH compulsory installation in Beijing is that solar water heating systems shall be installed in all new buildings with different solar fraction.
2. Subsidy to peasants in rural area:
   The policy of giving subsidies to peasants in the rural areas of China was carried out for three years from 2009 to 2011. The peasant who has bought a solar water heater can get a subsidy from a special foundation supported by the Ministry of Finance of the PRC. The subsidy value equals 13% of the cost of SWH.
3. Demonstration projects and demonstration cities and counties for renewable energy application in buildings: (China Association of Building Energy Efficiency, 2013)
   From 2006, the Chinese government started many demonstration projects for renewable energy applications including solar heating and cooling; the important ones include "Demonstration Items for Renewable Energy Application in Buildings" (386 projects altogether from 2006 to 2008) and "the Demonstration Cities and Counties for Renewable Energy Application in Buildings" (74 cities and 170 counties from 2009 to 2011) supported by the Ministry of Finance and Ministry of Housing and Urban-Rural Development of PRC. The projects, cities, and counties that are approved for demonstration can get financial support from the Ministry of Finance and the Ministry of Housing and Urban-Rural Development of the

PRC. In 2012 demonstration items were raised at the provincial level in some western provinces such as Qinghai, Yunnan, etc., where rich solar resources became demonstration provinces for solar application and financial support has increased.

4. Demonstration counties and cities for application of new energy resources:
From 2010, some items supported by other departments of China's government, such as the National Energy Administration, the Ministry of Finance, and the Ministry of Agriculture of the PRC, etc., were started; the important ones are "Demonstration Items for Green Counties" (108 counties in 2010), "Green Small Towns," "New Energy Cities," etc., from 2011. Having government support, solar heating and cooling in China will develop faster in the future.

10.3 Solar thermal heating and cooling technologies

China has formed a technical support system for solar heating and cooling including technology for manufacturing selective coating and solar collectors, performance testing and quality supervision systems, technical codes for system design, and computer optimum software, etc.

10.3.1 Technology for production of coating and solar collector

Research efforts have been ongoing over the years since 1979; the first cylindrical single-cathode magnetron sputtering system for producing an Al-N/Al selective absorbing surface was completed by Tsinghua University in China in 1986. Thus, China has developed a technology with independent intellectual property rights for making coatings. As the range of solar absorption and infrared emittance of the Al-N/Al selective absorbing coating is 0.93−0.95 and 0.05−0.06, respectively, the technology's industrialization has led to solar evacuated-tube collectors having a higher quality and market share in China.

China is a latecomer to the research of medium- and high-temperature solar collectors, and earlier research efforts were limited to small parts and appliances, but the situation is now changing. A noteworthy achievement began with the eleventh Five-Year Guideline, which involved the subject of substantial support for developing medium-temperature solar collectors from 2005 to 2010. The completed evacuated-tube collector in the subject could reach an operating temperature of over 160 °C. The use of a sol-gel-coated antireflection glass tube showed an increased solar transmittance to 0.94 (air mass (AM) 1.5) and solar absorptance of up to 0.96. Infrared emittance is 0.05 at 180 °C. The coating has an estimated durability of 25 years in a vacuum, and the glass tube is likely to last up to 15 years proved by 2500 h (400 °C) durability experiments. Figure 10.9 shows the efficiency curve of the collector; the value of the efficiency equals 0.52 when the reduced temperature difference is 0.13, so it has the better thermal performance when working in higher temperatures. Now, the medium-temperature solar collector has been produced by Linuo Paradigma Co., Ltd in Jinan, Shandong, and applied in some projects.

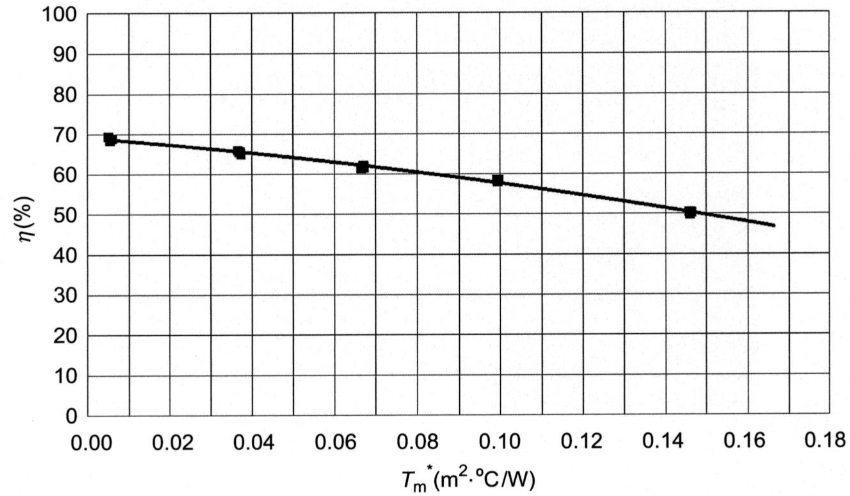

Figure 10.9 Efficiency curve of the medium-temperature solar collector developed by Tsinghua University.

10.3.2 National standards and quality supervision

China has established a better product quality supervision system including national standards for specification and test methods of systems, collectors, evacuated tubes, material, and main parts of the system, etc., setting up national testing centers to carry out national sampling, testing, and product certification, etc.

The following are China's main national standards: GB/T 18708, "Test methods for thermal performance of domestic solar water heating system," GB/T 19141, "Specification of domestic solar water heating systems," GB/T 4271, "Test methods for the thermal performance of solar collectors," GB/T 17581, "Evacuated tube solar collectors," GB/T 6424, "Flat-plate solar collectors," GB/T 17049, "All-glass evacuated solar collector tubes," GB/T 19775, "Glass-metal sealed heat-pipe evacuated solar collector tubes," GB/T 25969, "Specification of material selecting for main parts of domestic solar water heating system," and GB 26969, "Minimum allowable values of energy efficiency and energy efficiency grades for domestic solar water heating systems."

There are three national testing centers for quality supervision of solar water heating systems in Beijing, Wuhan, and Kunming. The National Center for Quality Supervision and Testing of Solar Heating Systems (Beijing) is in the China Academy of Building Research (CABR); two other National Centers (Wuhan and Kunming) are in the Hubei Provincial Supervision and Inspection Institute of Product Quality and Solar Energy Research Institute of Yunnan Normal University, respectively. Figure 10.10 shows the National Center for Quality Supervision and Testing of Solar Heating Systems (Beijing). Since 2004, the National Centers carried out yearly national sampling testing of domestic solar water heating systems, which is the task from the General Administration of Quality Supervision, Inspection and Quarantine

Figure 10.10 National Center for Quality Supervision and Testing of Solar Heating Systems (Beijing).

of the PRC every year, and it has promoted enterprises paying more attention to quality control and technological progress.

As in Europe, China also has its own certification for solar thermal products. Nowadays in China there are four main certifications for solar thermal products: CQC certification mark from the China Quality Certification Center, the Golden Sun certification mark from the China General Certification Center (CGC), the Ten Ring certification mark from the China Environmental United (Beijing) Certification Center Co., Ltd (CEC), which is mainly tied to the performance and environmental impact of solar thermal products, and the CABR certification mark from the CABR (Certification), which is mainly for product quality and performance integrated in buildings. Most products made in larger solar thermal enterprises of China passed above certification, which gives market competitive edges for the enterprises.

10.3.3 Technical codes and system design

Currently in China there are four national standards for engineering construction on solar heating and cooling that were jointly issued by the Ministry of Housing and Urban-Rural Development and General Administration of Quality Supervision, Inspection and Quarantine of the PRC. They are three mandatory standards: GB 50364-2005, "Technical code for solar water heating system of civil buildings," GB 50495-2009, "Technical code for solar heating system," and GB 50787-2012, "Technical code for solar air conditioning system of civil buildings," and two voluntary standards: GB/T 50604-2010, "Evaluation standard for solar water heating system of civil buildings," and GB/T 50801, "Evaluation standard for application of renewable energy in buildings," which includes evaluation of solar thermal applications in buildings.

According to these technical codes some important design handbooks on solar heating and cooling have been published in China in these years, such as "Technical guidebook for solar water heating system of civil buildings," Zheng Ruicheng, China Chemistry Industry Press, 2006, "Technical handbook for solar heat supply & space heating," He Zinian, Zhu Dunzhi, China Chemistry Industry Press, 2009, and "Technical handbook for solar heating," Zheng Ruicheng, Lu Bin, Li Zhong, He Tao, China Architecture & Building Press, 2012, etc.

Some important design parameters and computation methods are given in the handbooks, such as an experience formula for evaluation of the highest water temperature in a water storage pool when seasonal heat storage is over, evaluation methods for energy-saving effects and environmental effects, etc., so the engineers can understand very directly the design measures from codes and handbooks. After the publication of these technical codes and handbooks, many training classes were conducted to teach various engineers and technicians, so it has strongly promoted the raising of the design engineer's technical level.

10.3.4 Computer software and performance monitor

As computer software are very useful design tools for engineers, some of China's own computer software has been developed in recent years. Among them, "Design Software for Solar Heating and Cooling System," developed by the CABR is a better one, and it was one of the tasks of China's national research projects. The software can be used for design of solar water heating systems, solar heating combisystems, and solar cooling systems. There are three databases and four function modules in the software.

The three databases include a weather database, a performance parameters database of the solar collectors, which can update as new test results appear from the National Center for Quality Supervision and Testing of Solar Heating Systems (Beijing), and the graphics library of node—structure drawings for collectors integrated in buildings. Four functional modules are load calculation module, system-type selection and design module, system resistance calculation module, and analysis module for energy efficiency and environment protection effect of the system.

Solar heating and cooling has better energy-saving effects and should get financial support from the contract energy management, etc., to solve increasing investment, but this needs real energy-saving quantities of the systems. Therefore, the technology for system performance monitors developed very fast in recent years in China, and many projects have installed monitor systems that have set up transducers for testing temperature, flow rates, etc., and operation control equipment. Monitoring data such as temperature, flow rates, and solar irradiance, etc., can be transferred to a control center for optimum operation and to a long-distance computer by the Internet for calculating heat gain, etc., energy-saving effects. Figure 10.11 shows the monitor display screen of a solar cooling system.

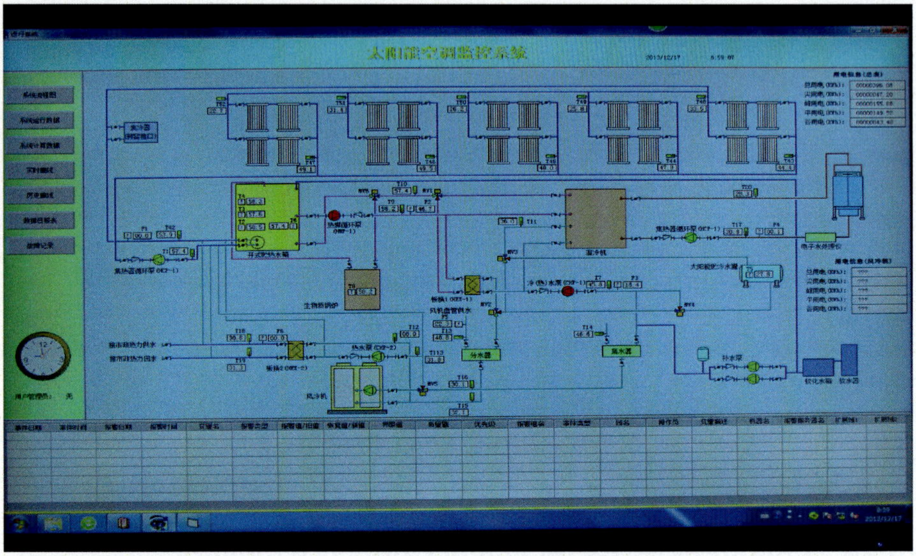

Figure 10.11 The monitor display screen of a solar cooling system.

10.4 Case studies of particular solar thermal installations in China

Introduced next are three case studies for solar thermal installations in China: a solar water heating system in the Din Xiu Xin Yuan Residential Quarter of Tianjin City (No. 1), solar heating combisystems with short-term heat storage in the peasant houses at Guajiayu Village of the Pinggu District in Beijing (No. 2), and a solar cooling system in the laboratory of the Institute of Built Environment and Energy Efficiency, CABR, in the Shunyi District of Beijing (No. 3).

10.4.1 Solar water heating system (No. 1)

This solar water heating system is for hot water supply to residents who live in 26 tall residential buildings of the Din Xiu Xin Yuan Residential Quarter of Tianjin City. The system should supply 80 L, 50 °C hot water to each family every day. The solar collector used in the system is a heat pipe evacuated-tube collector, and the total collector area of the system is 3148.1 m^2, which is installed on the roofs of 26 tall residential buildings. The slope angle of the collector installation is 39°. The system type is closed and the indirect system uses freeze-resistant liquid. One 80 L water storage tank is installed in the washroom of every family and one 1.5 kW electric heater as auxiliary heat source setting in the water tank. Figure 10.12 shows the solar collectors that are installed on the roofs of residential buildings in the Din Xiu Xin Yuan Quarter.

In November 2010, the National Center for Quality Supervision and Testing of Solar Heating Systems (Beijing) tested the thermal performance of the system that

Figure 10.12 Solar collectors that are installed on the roofs of residential buildings in the Din Xiu Xin Yuan Quarter.

is installed in residential building No. 23, which is 11 stories and has 44 families. Total solar collector area in this building is 59.08 m^2. Testing done for 4 days having different solar daily irradiation, and solar daily irradiation is $H < 8$ MJ/m^2, $8 \leq H < 13$ MJ/m^2, $13 \leq H < 18$ MJ/m^2, and $H \geq 18$ MJ/m^2, respectively, in these 4 days. From the testing data, the energy saving effect of the system in one year can be computed and analyzed.

10.4.2 Solar heating combisystem (No. 2) (Zheng et al., 2012)

This solar heating combisystem is in Guajiayu Village of the Pinggu District in Beijing. There are 71 houses in the village in which solar heating combisystems have been installed. Figure 10.13 shows this village. The range of building area of each house where installed solar heating system is 90 ~ 210 m^2 and the solar collector areas and volumes of water tanks in different houses range 14 ~ 60 m^2 and 0.3 ~ 1.5 m^3, respectively. The space heating type is low-temperature radiant floor heating. One house that installed 24.21 m^2 flat-plate collectors and uses 1.5 m^3 volumes of heat storage water tank was tested by the National Center for Quality Supervision and Testing of Solar Heating Systems (Beijing).

The testing was done during the entire winter of 2005 to 2006, and average room temperature was 12 °C during testing in winter when no other heating energy source assisted. On a sunny day of January the room temperature was between 10 and 15 °C, which can meet the comfort requirements of the winter room temperature for peasants now in China.

10.4.3 Solar cooling system (No. 3)

This solar cooling system is in the laboratory of Institute of Built Environment and Energy Efficiency, CABR, in the Shunyi District of Beijing. The total building area

Figure 10.13 Solar heating combisystems in Guajiayu Village of the Pinggu District in Beijing.

of the laboratory for solar air conditioning is 1850 m^2. U-type glass–metal evacuated-tube solar collectors measuring 524 m^2 are installed on the roof of the laboratory. The slope angle of the collector is 10°. Figure 10.14 shows the solar collectors of the system. The efficiency equation of the collector based on aperture area is $\eta = 0.732 - 2.371\ T_i$. There is one 176 kW lithium bromide absorption refrigerating machine, one 15 m^3 hot water storage tank, one 8 m^3 cold water storage tank, and one monitoring system in the system. The auxiliary heat source of the system is one 232 kW biomass boiler. The lithium bromide absorption refrigerating machine has a very wide range of working temperature from 70 to 95 °C, and the COP of the machine can reach 0.7 when the input hot water temperature from the solar collector is at 70 °C.

In summer 2013, the National Center for Quality Supervision and Testing of Solar Heating Systems (Beijing) tested this system. Testing was done for 4 days, from which different solar daily irradiation and solar daily irradiation was $H < 8$ MJ/m^2, $8 \le H < 12$ MJ/m^2, $12 \le H < 16$ MJ/m^2, $H \ge 16$ MJ/m^2, respectively, in these 4 days. From the testing data, the energy-saving effect of the system for the whole summer can be computed and was analyzed according to the methods in GB/T 50801.

10.4.4 Energy-saving effects for these three projects

Energy-saving effects of the above three projects are shown in Table 10.1. The effects are given according to testing results. The result of No. 2 is under the condition of 12 °C average room temperature in winter, and the result of No. 3 is under the condition of 26 °C average room temperature in summer. The heat value in Ton of Standard Coal Equivalent (TCE) units is 29.308 MJ/kg, and the CO_2 emission factor for 1 kg TCE is 2.47 kg CO_2/kg TCE.

Figure 10.14 Solar cooling system in the Shunyi District of Beijing.

Table 10.1 **Energy saving effects of three projects**

Project	Solar fraction %	Energy-saving TCE	Decreasing emission of CO_2 T
No. 1 (in one year)	70	594	1466
No. 2 (in one year)	100	1112	2746
No. 3 (in one summer)	83	16.7	41.3

10.5 R&D needs and future developing trends (Yin, 2008)

10.5.1 R&D needs for solar thermal heating and cooling

There are two main aspects in R&D needs for solar thermal heating and cooling in China, R&D for new products and R&D for technology of system integration.

1. R&D needs for new products
   As flat-plate solar collectors meet better the requirements of building integration, so the application of the flat-plate solar collector is increasing in the cities of China. However, in the past, most copper absorber plates with selective coating had to be imported from foreign companies, such as TINOX GmbH etc., so the higher cost influenced the capacity for market competition of flat-plate solar collectors. Therefore, an important R&D task is China's own production technology for making copper absorber plates with selective coating.

Solar air heating systems can be used in buildings for space heating and also process heat applications, e.g., for crop drying, etc. However, in the past there were fewer solar air collectors selected for use in China. Therefore, R&D for high-quality solar air collectors is another urgent need for China.

Nowadays, the auxiliary heat source in many solar heating systems is electricity that is generated using coal and results in air pollution in China. Therefore, the auxiliary heat source in solar heating systems should be changed from electricity to gas, etc., clean primary energy and R&D is needed for gas boilers that can adjust heating capacity to supply hot water having suitable working temperatures according to water temperature input from solar heating systems.

2. R&D needs for technology of system integration

   Technology of system integration includes two meanings: solar collectors integrated in buildings and system optimum design. Safety is the important factor of solar collectors integrated in buildings, and system optimum design will influence the energy-saving effects of the system.

   R&D needs for solar collectors integrated in buildings include the following aspects: construction technology for solar collectors integrated in buildings that is suitable to the character of China's building models and building materials; the series of fixing components that can be used when installing solar collectors on roofs, balconies, and walls; architectural design technology for safety guarantee, etc.

   R&D needs for system optimum design and operation include the following technologies: dynamic simulation for thermal performance and operational features of solar collector systems, suitable short-term and seasonal heat storage technology, intelligent control technology for system optimum operation, water-saving technology for solar water heating systems, optimum design for large-scale solar heating combisystems, comprehensive utilization for solar heating and cooling over the entire year, comprehensive utilization for solar energy, and other renewable energy sources such as geothermal energy and biomass energy, etc.

10.5.2 Future developing trends for markets and applications

1. Market

   The market for solar thermal heating and cooling in China has three developing trends: from the market in small cities and the countryside changing to the market in the large and middle-sized cities, from the market in eastern areas having high economic level changing to the market in western areas where people are poorer but have better solar resources, from a market of agent sales changing to the market sold by system integrators.

2. Applications

   Applications for solar thermal heating and cooling in China have two developing trends: close-coupled domestic solar water heating systems changing to remote storage solar water heating systems and central solar water heating systems, and single applications of solar water heating systems changing to comprehensive application of solar heating combisystems and solar cooling systems.

3. Strategic goals for development

   For better and faster development of solar thermal heating and cooling, and giving more attention to climate change and environmental protection, research for developing the roadmap for solar thermal in 2020, 2030, and 2050 has been completed in China. The roadmap research gives China's strategic goals for development of solar thermal in

Table 10.2 Strategic goals for development of solar heating and cooling at the usual scenario

Year	Solar hot-water GWth/year	Solar heating combisystem GWth/year	Solar cooling GWth/year
2020	423	25	20
2030	515	64	51
2050	629	169	135

Table 10.3 Strategic goals for development of solar heating and cooling at positive scenarios

Year	Solar hot-water GWth/year	Solar heating combisystem GWth/year	Solar cooling GWth/year
2020	577	45	36
2030	776	139	113
2050	1045	446	363

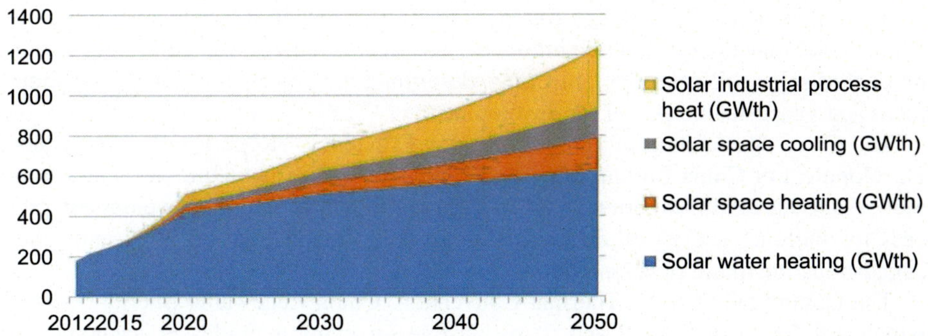

Figure 10.15 Roadmap vision for solar thermal based on the usual scenario in China.

the cases of the usual and positive scenarios; the strategic goals for development of solar heating and cooling based on usual and positive scenarios are shown in Tables 10.2 and 10.3. Figures 10.15 and 10.16 show the roadmap vision for solar thermal applications in China based on usual and positive scenarios, respectively.

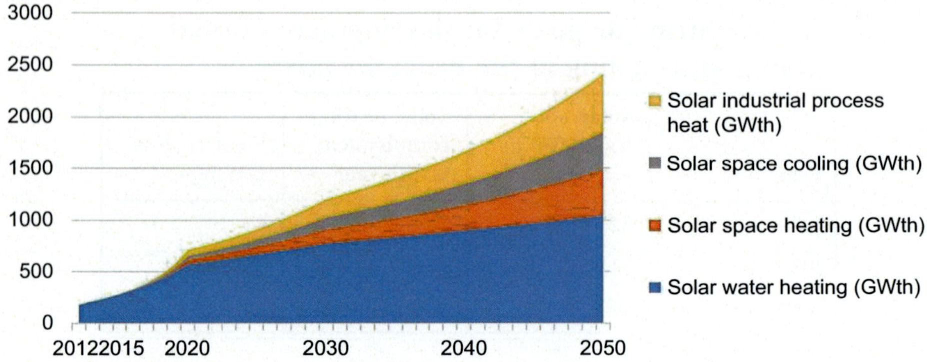

Figure 10.16 Roadmap vision for solar thermal based on positive scenarios in China.

Sources of further information

Sources of further information and advice in this chapter are from some key books, professional bodies, research groups, and websites. The main ones are as follows:

Xu Wei, et al., "Report on China Solar Energy in Buildings," which summarizes China's industrial policy, national standards and codes, market situation, etc., for solar applications, introducing characteristic technology, construction experience, and typical projects of solar heating and cooling, etc., and giving the trends developing in the future.

Zheng Ruicheng, et al., "Technical Handbook for Solar Heating," which discusses the methods for load computation, system design and installation, system monitoring and evaluation for energy-saving effects of the solar heating systems, etc., giving suitable design parameters and some actual engineering projects for solar heating.

Yin Zhiqiang, et al., "Strategy Research for Development of Renewable Energy in China Vol. Solar Energy," which introduces the solar resources and technology development of solar collectors, etc., in China, describing the objects for industry development and guarantee measures for development.

The China Association of Building Energy Efficiency, "Report on the Status and Development of China Building Energy Efficiency (2012)," which reviews and summarizes the successful experiences and existing problems during the development process for building energy efficiency including solar applications in 2012 and gives suggestions for future development.

The Committee of Solar Thermal Conversion, China Renewable Energy Society, which was founded in 1979 and is responsible for many activities in academic exchange, technology training, etc., of solar heating and cooling. It completes the Roadmap Research of China Solar Thermal Development in 2020, 2030, and 2050. Its website is www.cres.org.cn.

The Committee of Solar Thermal Application, China Rural Energy Industry Association, which was founded in 1994 and is responsible for communication between enterprises and the government, promoting market normalization and enterprise self-discipline, etc. It also gives the yearly report on the development of China's solar thermal industry.

The China National Renewable Energy Center, which was founded in 2011 and is responsible for research on strategy development, making plans and policy for renewable energy, etc. Its website is www.cnrec.org.cn.

References

China Association of Building Energy Efficiency, 2013. Report on the Status and Development of China Building Energy Efficiency (2012). China Architecture & Building Press, Beijing.

Xu, Wei, Zheng, Ruicheng, Lu, Bin, 2009. Report on China Solar Energy in Building. China Architecture & Building Press, Beijing.

Yin, Zhiqiang, 2008. Strategy Research for Development of Renewable Energy in China Vol. Solar Energy. China Electrical Power Press, Beijing.

Zheng, Ruicheng, Lu, Bin, Li, Zhong, He, Tao, 2012. Technical Handbook for Solar Heating. China Architecture & Building Press, Beijing.

Heat generation from biomass in Sweden

11

Claes Tullin
SP Technical Research Institute of Sweden, Borås, Sweden

11.1 Introduction

This chapter is intended to give an overview of the development of biomass use for heat (and power) production in Sweden, to highlight the important implementation of supporting policies, to give examples of the present use, and finally to give an outlook for the future.

First, a background is presented to describe the Swedish context as a forest nation with a large forest industry sector. A brief historic overview is given, and the past decade's development and use by different sectors of biomass are presented. Subsequently, some attention is given to show the important effect of policy drivers for the market development of the use of biomass (and waste) to become the single largest energy source in Sweden.

The potential of using biomass and waste in practice in large as well as in small scale is illustrated by a number of case studies:

- Small-scale pellet combustion
- Small-scale district heating
- Large-scale combined heat and power (CHP) production
- Large-scale waste to energy
- Energy combines

The chapter concludes with describing the research and development (R&D) needs and future trends in technological development, markets, and applications. Finally, some suggestions for further reading are given.

11.2 Background

Sweden is a relatively large country (450,000 km^2) with low population density (9.6 million inhabitants). With about 280,000 km^2 in total, forests cover well over half of the land surface (SLU, 2015).

The forest industry is one of Sweden's most important industrial branches, corresponding to 10–12% of Swedish industrial employment, revenue, and exports. Exports include pulp and paper, of which about 90% is exported, and sawn timber with about 75% on export. In a global context this means that Sweden is the second largest exporter of paper, pulp, and sawn timber in the world (Forest Industries, 2013).

Renewable Heating and Cooling. http://dx.doi.org/10.1016/B978-1-78242-213-6.00011-4

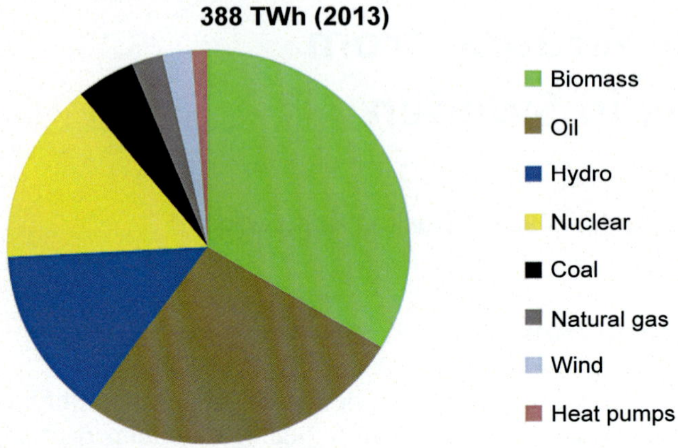

Figure 11.1 Final energy use in Sweden 2013.
Source: Data compiled by Svebio (2015) based on statistics from the Swedish Energy Agency.

The number of employees is about 60,000. Including subcontractors, etc., the number of employees is around 200,000. Bioenergy has always been an integrated part of the forest industry, but, during the past few decades, the forest industry production of bioenergy for both internal and external use (as heat, power, and fuels) has increased substantially to become an increasingly important part of the business.

Wood was for long the main source for heating in houses, mostly in very inefficient (typically less than 10% efficiency) open fireplaces. In the eighteenth century, the lack of wood became severe as a consequence of iron manufacturing and the use of timber in Sweden and Europe not being balanced by management regarding harvesting and replantation. In the second half of the eighteenth century, the need for more energy-efficient heating systems resulted in the development of the Swedish tile stove with much higher efficiencies (Cronstedt, 1767). In modern tile stoves the efficiencies can be close to 90%. The tiled stoves were subsequently replaced by central heating systems (boilers) using wood (and coal), but after the second world war oil became the dominating fuel for heating in Sweden. The massive introduction of nuclear power in the 1970s resulted in low prices for electricity and heating based on direct electricity heating as well as electric boilers becoming common. As a consequence of the oil crises and, more recently, the rise of policies supporting the introduction of renewable energy sources, the biomass share in the energy system has increased to become the dominating energy source today with about one-third of the energy supplied (Figure 11.1). In the past 30 years the use of biofuels and peat[1] has almost tripled and is expected to continue to increase (Figure 11.2).

It is important to note that there is no apparent conflict between an increased use of bioenergy and climate mitigation as long as annual harvest levels do not exceed annual growth on a landscape or national level. Despite the increased use of bioenergy in

[1] The supply of peat for heat and electricity production in 2011 was about 3.1 TWh or 2% of the bioenergy.

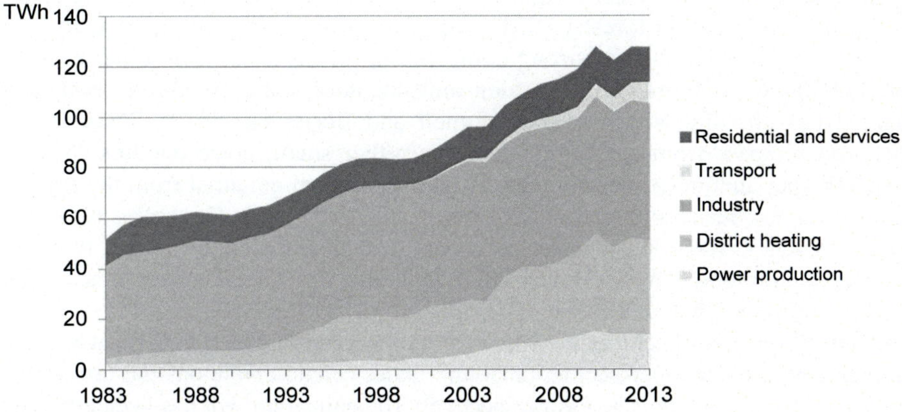

Figure 11.2 Use of biofuels and peat by sector 1980–2013.
Source: Based on the statistics from the Swedish Energy Agency.

Sweden, the Swedish forests have shown a net growth throughout the twentieth century. In Figure 11.3, the difference between gross felling and annual increment is shown for the past 60 years (Swedish Forest Agency, 2014). Furthermore, in a study of the future role of Swedish forestry through the year 2105, Lundmark et al. (2014) concludes that with present strategies long-term climate change mitigation corresponds to more than 60 million tons of CO_2 annually, equivalent to current CO_2 emissions in Sweden. Using more intensive silvicultural methods can increase the long-term mitigation effect by another 40 million tons. Thus, active management of forests can provide renewable resources at the same time as the forests grow and capture CO_2 from the atmosphere.

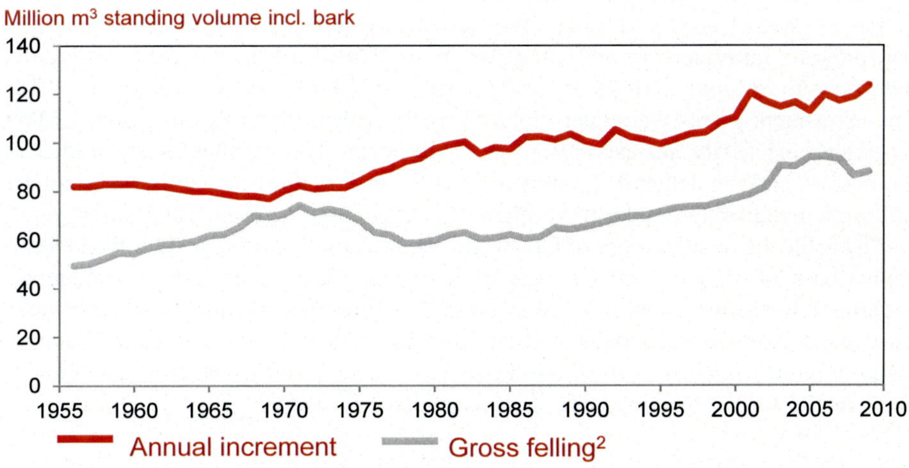

Figure 11.3 Annual gross fellings and increment in the Swedish forests. The numbers are 5-year averages (Swedish Forest Agency, 2014).

In addition to forest fuels, significant efforts have been made in Sweden to introduce short-rotation forestry (SRF) with fast-growing tree species and rotation in less than 20 years. Due to strong competition, for instance, from other biomass the 33,000 km^2 of farmland in Sweden only includes some 10,000–11,000 ha of planted fast-growing willow (*Salix*) (Egnell and Börjesson, 2012). Thus, use of SRF and biomass from agriculture today is rather small, corresponding to some 2–3 TWh per annum (Andersson, 2012). The theoretical potential from the agriculture sector has been estimated at 23 TWh in 2020 (Swedish Government Official Reports (SOU), 2007:36). The technoeconomical potential is of course less, and the present ambition is 6–8 TWh by 2020, including *Salix*, reed canary grass, straw, etc. (Energimyndigheten, 2011).

Although peat is sometimes classified as "slowly renewable," it is not considered a biofuel from a climate perspective (although under certain conditions the use of peat is positive from a greenhouse gas perspective). The annual growth has been estimated at 18–20 TWh in Sweden (SOU, 2002:100), but the present use does not correspond to more than about 3 TWh or some 2% of the combined supplied energy of biomass and peat.

11.3 Market development

11.3.1 Policies and market drivers

The introduction of the CO_2 tax in 1991 was very important in phasing out fossil fuels for CO_2-neutral alternatives in Sweden. The tax is differentiated between industry[2] and other sectors (households, service, etc.). The tax has been raised strongly for the nonindustrial sector on a number of occasions and now accumulates to some 115 €/ton CO_2 (Figure 11.4). For industry, the rather steep increase in 2015 is expected to be an important driver for the replacement of fossil fuels.

Power production from biomass was previously supported by an investment support program introduced in 1991. Support up to 30% of the investments was granted with the aim to build CHP plants and produce 1 TWh biopower (Andersson, 2012). The investment grants were later replaced by the green certificates in which 1 MWh renewable electricity corresponds to one certificate. The certificates are traded on the market, and the demand is maintained as the electricity distributors are stipulated to have a certain share of certified renewable electricity that increases year by year.

To enable the increased use of bioenergy, the governmental support for R&D has of course been of vital importance. Academic research has developed knowledge and expertise and has been a base for educational programs of good quality and relevance. To enhance knowledge transfer and the implementation of new knowledge, applied R&D programs in close collaboration between industry and research institutions has been crucial for the development of a vital bioenergy sector.

[2] From 2013, the industrial CO_2 tax does not include industry within the Emissions Trading Scheme (ETS).

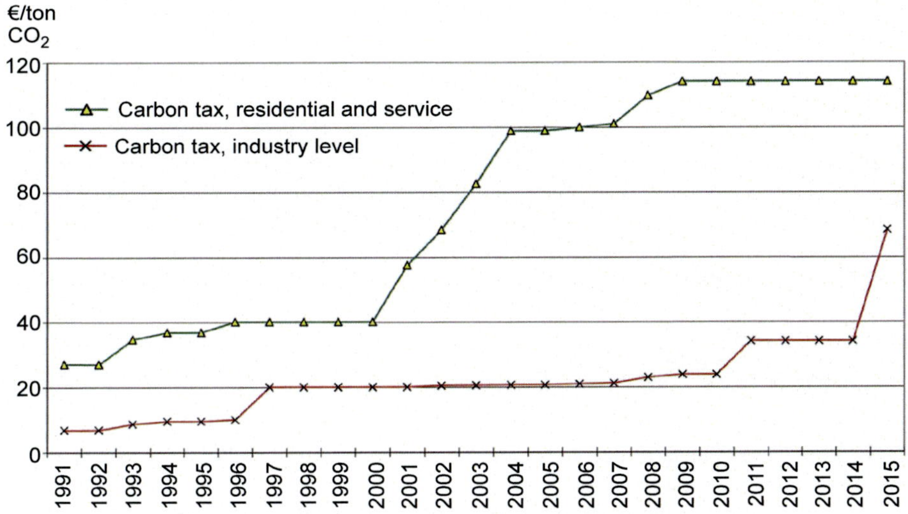

Figure 11.4 Development of the carbon tax in Sweden (KSLA, 2013).

11.3.2 District heating and the residential sector

In the residential sector, oil has largely been replaced by renewable alternatives, and today district heating, biomass, and heat pumps are dominating new installations (Figure 11.5).

The use of district heating has increased about four times during the past 40 years, mainly for use in the residential and service sectors but also in industry (Figure 11.6). Most of the customers are found in multifamily residential housing but also single-family homes are connected to a district heating grid in some areas. In the beginning

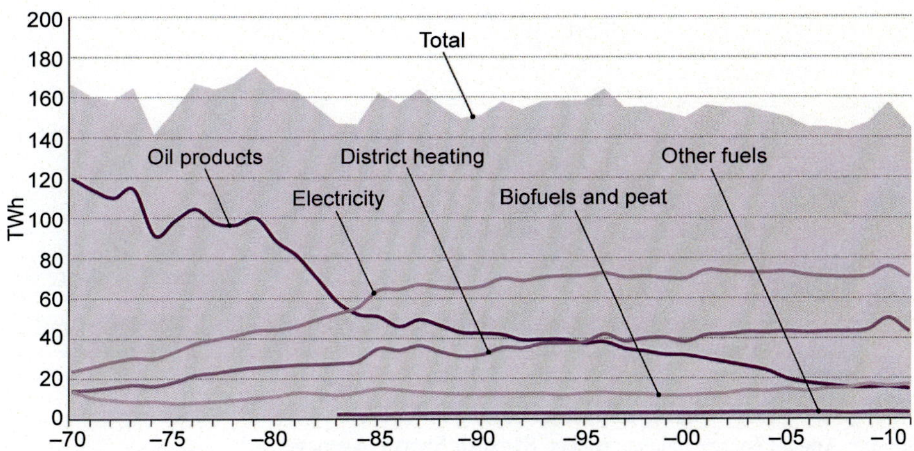

Figure 11.5 Energy use in the residential sector.
Source: Annual energy balance sheets. Swedish Energy Agency.

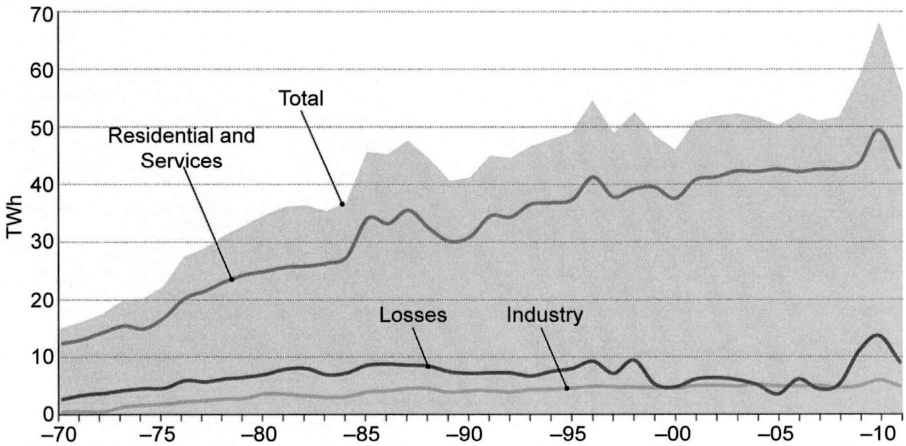

Figure 11.6 Development of district heating in Sweden. The peak in 2010 was due to an unusually cold winter season.
Source: Annual energy balance sheets. Swedish Energy Agency.

of the 1970s, the sector was dominated by the use of oil (Figure 11.7). As a consequence of the oil crises, other sources like coal, biomass, heat pumps, and use of waste heat were introduced as the system expanded and to replace oil. The introduction of a CO_2 tax on fossil fuels in 1991 gave room for a switch from fossil to renewable sources, with biomass dominating the development. In Figure 11.8, the present supply of district heating on different sources is shown.

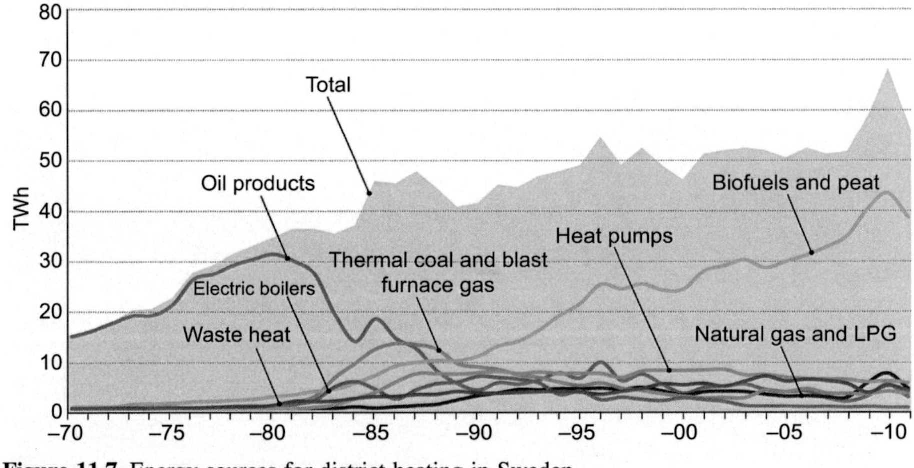

Figure 11.7 Energy sources for district heating in Sweden.
Source: Annual energy balance sheets. Swedish Energy Agency.

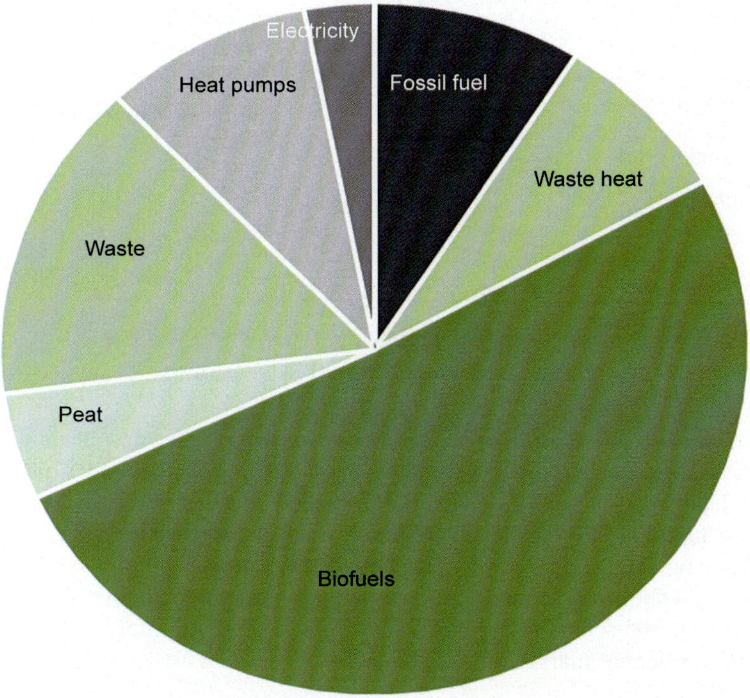

Figure 11.8 Supplied energy for heat production in district heating (Avfall Sverige, 2014).

The total number of heat plants using biomass, waste, and peat is around 500 (Andersson, 2014). Most of these plants use wood chips, residues from forest industries and forestry, or recycled wood. But many of the smaller ones also use pellets. Pellets and bio-oils are often used for top load and in summer.

Of these 500 heat plants, 91 also produce electricity (CHP). Almost all of Sweden's 290 municipalities have heat plants and district heating based on biomass. In many cases several communities in a municipality have district heating grids. Also the biggest cities, like Stockholm, get most of their district heating from biomass and waste.

11.3.3 Industrial heating

As mentioned in the Introduction, bioenergy is an integrated part of the forest industry used, for instance, in recovery boilers in the pulp and paper industry and for drying in saw mills. The increase in the CO_2 tax 2015 will lead to replacement of fossil fuels with alternatives such as biomass also in other industry sectors.

11.3.4 Pellets

As a consequence of the introduction of the CO_2 tax, it became very profitable to replace oil burners with pellet burners. Pellet burners for domestic and small-scale use were

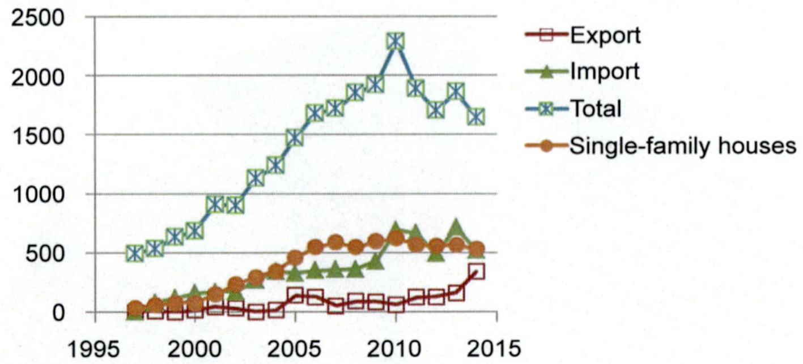

Figure 11.9 Pellet production and use in Sweden (kton) (Pelletsförbundet, 2015).

developed in Sweden to replace oil burners in existing boiler and heating systems. The market expanded dramatically in the early 1990s (Figure 11.9) but has now come to a halt. Important for the market introduction of pellet burners was a voluntary so-called P-marking (SP Technical Research Institute of Sweden (SP), 2015) with requirements on safety, emissions, efficiency, and function.

Of the pellet deliveries around 30% goes to large-scale users (above 2 MW), around one-third to middle-scale users (25 kW–2 MW), and a little more than 40% to the small-scale residential market.

11.3.5 Waste

In Sweden, a number of policies have been introduced to avoid and reduce waste and to optimize the utilization of waste. In 2000, a landfilling tax was introduced that since then has been increased on several occasions. A ban on landfilling of combustible waste was introduced in 2002 followed by a ban on landfilling of organic waste in 2005. This has resulted in minimizing landfilling despite the fact that the total amount of waste has increased (Figure 11.10(a) and (b)) as the capacity for incineration, biological treatment, and recycling has increased.

Substantial applied R&D efforts have also been made to enable the use of demolition wood in conventional biomass boilers (Värmeforsk, 2014). The work has been successful, and today a number of plants are using demolition wood as part of the fuel mix. Co-combustion of sludge with biomass has also been devoted larger R&D efforts and also been proven to have positive effects—for instance, when co-combusted with waste wood (Åmand et al., 2006a,b).

As a result of the introduced policies, the Swedish capacity for energy recovery more than doubled during the period 2000 to 2008 (Waste Refinery, 2013). The interest to invest in CHP for energy recovery has been maintained, and, if all planned investments are realized, the capacity would increase to some 7 Mton per annum in 2017 (Waste Refinery, 2013). The Swedish waste has accumulated to about

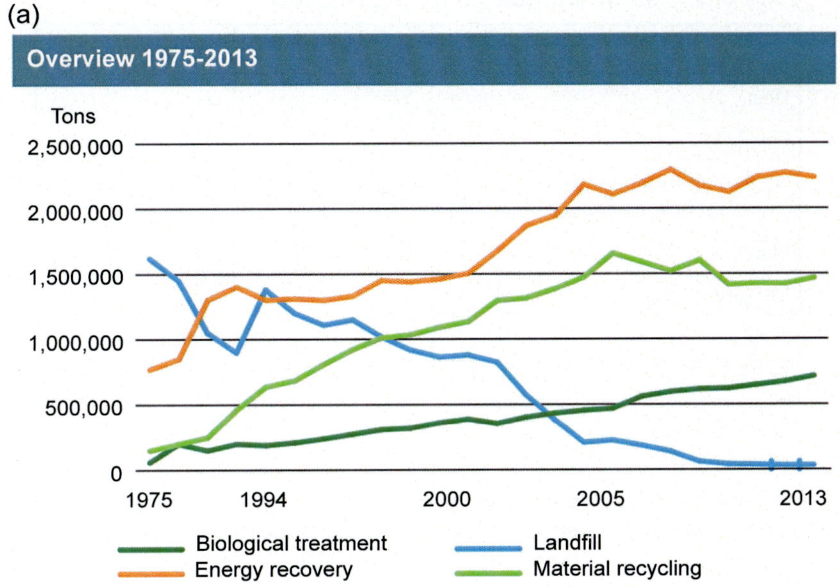

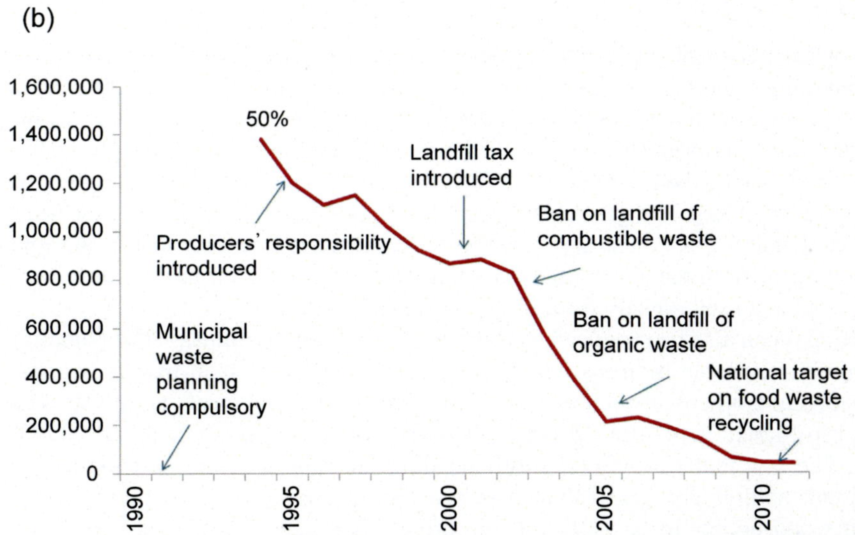

Figure 11.10 (a) Development of waste management in Sweden over the past decades (Avfall Sverige, 2014). (b) The development of landfilling (tons) as a consequence of introduced policies (Avfall Sverige, 2014).

4.5 Mton and the rest is imported. In this way, the very efficient Swedish energy recovery plants also contribute to reduce landfilling in the EU. It should be emphasized that from a climate perspective this makes a lot of sense despite the increased transports as long as the waste management system in Europe has limited capacity. The main reason

Table 11.1 Use of biogas 2013 in Sweden

	Usage (GWh)	Share (%)
Heat	521	31
Power	46	3
Upgrading to biofuel *Including LBG (liquid biogas)*	907 *33.5*	54
Flare	186	11
Missing data	26	2
Total	**1686**[a]	**100**

[a]The annual potential from anaerobic digestion in Sweden is estimated at 3–4 TWh (Swedish Energy Agency, 2010).

is that landfilling contributes strongly to greenhouse gas emissions[3] and that both the power and heat produced can be utilized very efficiently in Sweden.

11.3.6 Biogas

Biological treatment (anaerobic digestion) and biogas production are also becoming increasingly important (Table 11.1). Two trends are that the total production increases and an increasing fraction is upgraded for use as transportation fuel. In 2013, the biogas production accumulated to almost 1.7 TWh in 204 plants and 60 landfills. The content of methane is usually between 60% and 70%, depending on the substrate, with the balance being mostly CO_2. The gas can be upgraded to biofuel or used as such for heat and power. Landfill gas is also harvested to avoid methane emission to the atmosphere. As nitrogen (from the air) constitutes a large part of the landfill gas it is normally not upgraded and used directly for heat and power production.

Biogas can also be produced by thermal gasification of biomass. The potential depends on available biomass and is naturally much larger than biogas from waste streams. In Gothenburg, a demonstration plant for thermal gasification was inaugurated in March 2014—the Gobigas plant. The plant will produce about 20 MW of gas. The raw material will be forest residues, and the gas will be supplied to the regional natural gas grid (35 bar) in the Gothenburg area. The plant started up with wood pellets in late 2014. If the project is successful, the second stage will contribute another 80–100 MW of gas, which would correspond to about 1 TWh biogas in 2020. About 65% of the raw material is converted to gas on an energy basis, and the overall efficiency of the plant is over 90%, when including heat recovery to district heating (Figure 11.11).

[3] Methane is produced in landfills by anaerobic digestion. As a greenhouse gas 1 kg of methane corresponds to 23 kg of carbon dioxide.

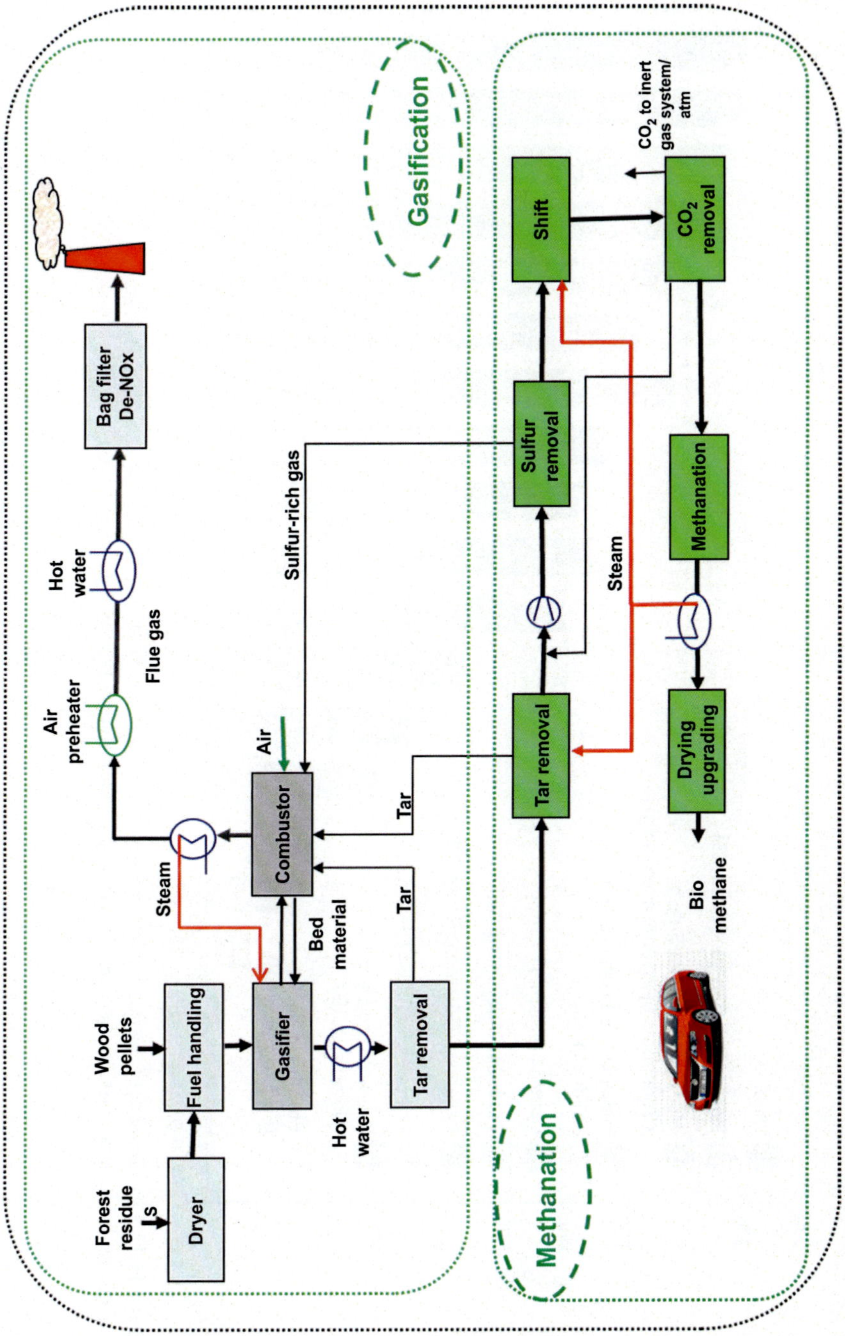

Figure 11.11 The 20 MW Gobigas gasification plant in Gothenburg.
Source: Image provided by Göteborg Energi.

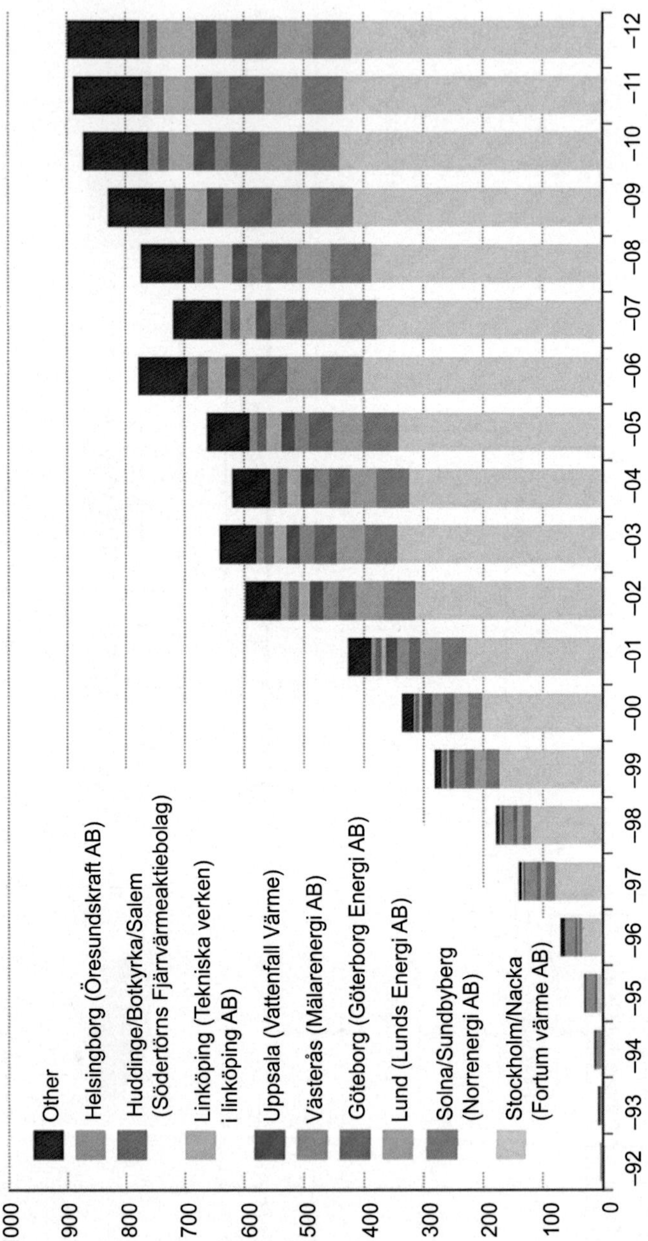

Figure 11.12 Development of district cooling in Sweden.
Source: Swedish District Heating Association, adapted by the Swedish Energy Agency.

11.3.7 District cooling

District cooling in Sweden is a fairly new phenomenon with the first plants inaugurated in the early 1990s. Today some 30 plants produce a district cooling equivalent to 0.9 TWh (Figure 11.12). The potential need for district cooling in Sweden is estimated to 2–5 TWh. District cooling is used in office buildings, shops, and hospitals and to cool industrial processes. Waste heat or lake water is used as a heat source in heat pumps to produce cool temperatures. It is also common to use cold bottom water from a lake or the sea, so-called free cooling.

11.4 Case studies of particular biomass installations

The following case studies are presented to highlight the use of biomass for heating in Sweden in small, medium, and large scale. High system efficiencies are obtained based on the use of heat in co-generation plants. In 2013, the number of biomass CHP units accumulated to 133 including some 42 industrial units. In addition, many units produce only heat on small and medium scales. Waste fuels will continue to be important in the future when challenges include higher steam data for co-generation. Finally, the transition to a bioeconomy provides new opportunities for highly resource-efficient energy combines.

11.4.1 Small-scale pellet combustion

Pellet burners were the first generation of combustion equipment developed for use of pellets at domestic scale (Table 11.2). Pellet burners were developed in Sweden to

Table 11.2 Case study of a pellet boiler

Parameter	
Site	Detached house
Type of technology	Pellet boiler
Fuel power input, nominal	10.9 kW
Nominal thermal capacity	10.0 kW
Pellet consumption at nominal load	2.3 kg/h
Boiler efficiency, nominal load	92%
Annual heat supply	13,300 kWh (1330 h at full load in average the first 3 years of operation)
Fuels	Wood pellets
Fuel moisture contents	<10%

The Pellet Handbook (2010).

replace oil burners in existing oil or multifuel boilers. With the introduction of the Swedish carbon dioxide tax in 1991, the conversion from oil to pellets became very profitable. Today, in principle, all domestic oil burners have been replaced with alternative and renewable technologies such as pellets.

11.4.2 Small-scale district heating—dry fuel

There are a number of smaller biomass boilers that have replaced oil boilers in smaller district heating networks. The boilers use wood chips or refined fuels, i.e., pellets or briquettes.

As an example of small scale installations Osby Parca has a container-based system (Eliasson, 2014). The system is fully automatic and include boiler and systems for fuel and ash handling and flue gas cleaning (see Figure 11.13 and Table 11.3).

11.4.3 Middle-size district heating—wet fuel

A middle-size heat central cover between 3 and 20 MWth and is suitable in heating systems for smaller to mid-size district heating grids, energy-intensive industries such as saw mills, and process industries or a small-scale CHP. It is possible to get hot water, saturated steam, or overheated steam according to customer requirements.

Figure 11.13 Small-scale biomass boiler for dry fuels.
Source: Photo provided by Osby Parca.

Table 11.3 Small-scale district heating plant using dry fuel—Summary of data

Parameter	
Site	Oskarshamn
Type of technology	Osby Parca, PB2—Fix grate for dry fuels, boiler with max 6 bar
Fuel power input, nominal	2 × 3 MW
Nominal thermal capacity	6 MW
Boiler efficiency, nominal load	94%
Annual heat supply	48 GWh
Fuels	Wood pellets
Fuel moisture contents	<10%
Flue gas cleaning	Multicyclones
In commercial operation	2012

Figure 11.14 Medium-scale biomass boiler for wet fuels.
Source: Photo provided by Jernforsen.

The boiler manufacturer takes care of the design and planning, manufacturing, assembling, and start-up of the plant. The plant is in fully automatic operation from the fuel feeding and combustion to ash removal and flue gas cleaning. The combustion system ensures low emission and high efficiency without impact on the availability and maintenance (Figure 11.14 and Table 11.4).

11.4.4 Large-scale CHP: one of the world's largest plants for CHP production under construction in Stockholm

With an annual production of 8300 GWh heat, 400 GWh cool, and 1500 GWh electricity, Fortum AB is one of the largest suppliers of heat and power in the Stockholm

Table 11.4 Middle-scale district heating plant using wet fuel—Summary of data (Jernforsen, 2014)

Parameter	
Site	Tönsberg, Norway
Type of technology	Jernforsen FF-furnace with moving grate for wet fuel Boilers 16 bar
Boiler power output, nominal	5 + 10 MWth
Nominal thermal capacity	15 MWth
Boiler efficiency, nominal load	90%
Annual heat supply	120 GWh
Fuels	Wood chips, bark, residue from forest harvest, residue from saw mills
Fuel moisture contents	35–55%
Flue gas cleaning	Multicyclones + ESP (Electrostatic precipitator)
In commercial operation	2012

area. As part of the development to carbon dioxide-free production, one of the largest boilers for biomass in the world is now under construction at the production site of Värtaverket in central Stockholm (Figure 11.15 and Table 11.5). The new biofueled CHP plant in Värtan is an important step in the development of a sustainable energy supply in Stockholm, Sweden, and the rest of Europe. The renewable energy will be obtained from waste products from the forestry industry such as chips, bark, branches, tops, and twigs. The boiler will consume about 3 million m^3 of fuel annually, or

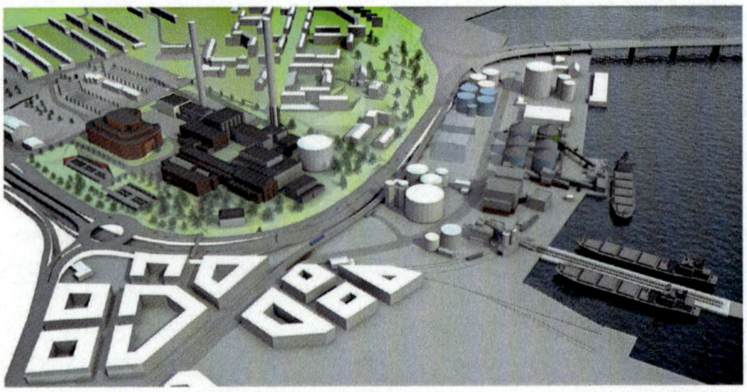

Figure 11.15 The new biomass combined heat and power plant under construction in Värtan, Stockholm.
Source: Image from Urban Design/Gottlieb Paludan.

Table 11.5 Data for the new combined heat and power plant in Värtan, Stockholm

Parameter	
Type of technology	CFB—circulating fluidized bed
Steam power, nominal	330 MW
Nominal thermal capacity	280 MW
Including flue gas condensation	80 MW
Nominal electric capacity	130 MW
Steam data, nominal load	560 °C/140 bar
Thermal efficiency, nominal load	91.5%
Electric efficiency, nominal load	40%
Annual heat supply	1700 GWh
Annual electricity generation	750 GWh
Fuels	Forestry and logging residues (tree tops, branches, bark, stumps, wood chips)
Ash handling	Primary objective is to recycle the ash to the forest
Optional fuels	Peat, coal
Fuel moisture contents	20–55%
Annual fuel demand	3 million m^3 forest residues
Flue gas cleaning	Textile filter
	Flue gas condensation
	SNCR including slip catalyst
Investment	Approximately 4 billion Swedish krona (SEK)
In commercial operation	2016

12,000 m^3 per day. The new boiler will have a steam power of 330 MW with an electrical output of 130 MW. The boiler technology is a circulating fluidized bed (CFB) delivered by Andritz, and the steam turbine is delivered by Doosan Skoda Power. The CFB technology ensures good fuel flexibility including different forest residues as well as peat and coal.

The fuel will be supplied by ships (60%), train (>30%), and trucks (<10%) preliminary from Sweden, Finland, the Baltic States, and Russia. To minimize emissions from the ships, they will be connected to the electrical grid in the harbor. Special concern has been taken regarding the unloading to ensure low emissions of dust and noise and during nighttime the operation will be run in "silent mode" at lower capacity. The fuel will be transported in a tunnel (10 × 8 m wide) to an underground storage

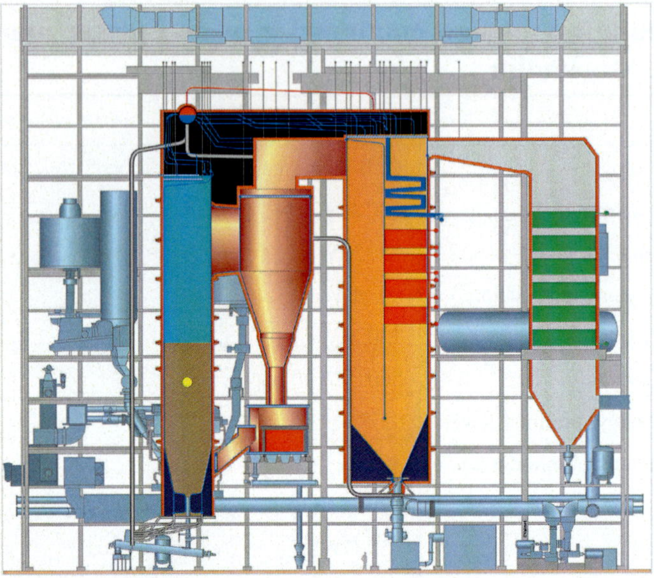

Figure 11.16 The new 167 MW fuel input circulating fluidized bed boiler in Västerås.
Source: Illustration by Valmet.

(rock cavity) of some 50,000 m^3 (corresponding to 4 days of operation). The bottom ash and fly ash will be transported in the tunnel in the opposite direction.

When the new plant is commissioned in 2016, 1700 GWh of heat and 750 GWh of electricity will be produced annually. That is enough to heat around 190,000 apartments. The produced renewable electricity will have a positive impact on the global climate by replacing electricity that would otherwise be produced from fossil fuels in the Nordic Region and in Europe. That means that the new biofueled CHP plant alone will reduce Europe's CO_2 emissions by 650,000 tons per year. The CO_2 emissions in Stockholm will be reduced by 126,000 tons per year—equivalent to the amount emitted by vehicles in Stockholm in 1.5 months.

11.4.5 Large-scale waste to energy

Valmet Power Oy has long experience in advanced CFB[4]—technology for waste to energy using solid recovered fuels (SRF). The size of the plants start at 50 MW fuel input, with a maximum size expected of 250 MW with the present technology. Main benefits of CFB combustion technology are high fuel flexibility (wide span in heating value, ash, and moisture contents), high combustion efficiency, and low emissions. Also the availability is high with reference data from SRF combustion, indicating that an availability of more than 97% is achievable (Luomaharju and Viljanen, 2013).

[4] CFB—Circulating Fluidized Bed.

Table 11.6 Data for the new circulating fluidized bed (CFB) waste-to-energy plant (Mälarenergi unit 6) in Västerås

Parameter	
Type of technology	CFB
Fuel power input, nominal	167 MW
Nominal thermal capacity	155 MW (From boiler to steam cycle) ca. 100 MW back pressure and 30 MW FGC (Mälarenergi website)
Nominal electric capacity	Ca. 50 MW
Steam data, nominal load	468 °C/74 bar
Thermal efficiency, nominal load	Boiler efficiency ~90%
Electric efficiency, nominal load	0.46 alpha value
Annual heat supply	860 GWh (Mälarenergi website)
Annual electricity generation	320 GWh (Mälarenergi website)
Fuels	Table 11.7 below
Annual fuel demand	Corresponding to 3 million m^3 forest residues
Flue gas cleaning	Alstom NID
Investment	Whole plant ~300 million euros (MEUR)
In commercial operation	2014

Valmet is delivering the so far largest boiler in the world for solid recovered fuel to Mälarenergi in Västerås (Figure 11.16). The fuel input will be 167 MW, corresponding to 60 tons of fuel per hour. The boiler is designed to burn household waste up to 70% and industrial waste up to 100% of heat input. Other fuels include wood, peat, and biomass (cf. Tables 11.6 and 11.7).

11.4.6 Energy combines

The demand for ethanol as biofuel is expected to increase strongly as a means to reduce greenhouse gas emissions. Sweden, where ethanol, fossil biofuel, and protein feed are imported and where cereals are normally exported, provides an interesting market for ethanol production. Furthermore, there is no conflict with food or feed production as the used farmland in Sweden is decreasing.

In Norrköping, Lantmännen Agroetanol AB is operating an ethanol production plant. The first production line was inaugurated in 2001 and the second in 2008. The ethanol production is integrated into the energy combine of Händelö in Norrköping (Figure 11.17 and Table 11.8), producing CHP from biomass and waste. Heat is fed to the district heating network of Norrköping, and process heat in the form of steam is delivered to the ethanol factory. This ensures very high system efficiencies.

Table 11.7 **Fuel properties**

Fuel	Moisture (%)	Lower heating value, LHV (MJ/kg)	Ash (weight %, dry)	Fraction (% LHV)
Recycled wood	20–40	10–14	2–7	0–70
Biofuel mix	25–55	8–13	1.2–5	0–70
Crushed peat briquettes	13.5–21	14.4–17.3	12–16	0–30
Sewage sludge	70–80	0.7–1.2	40–50	0–4

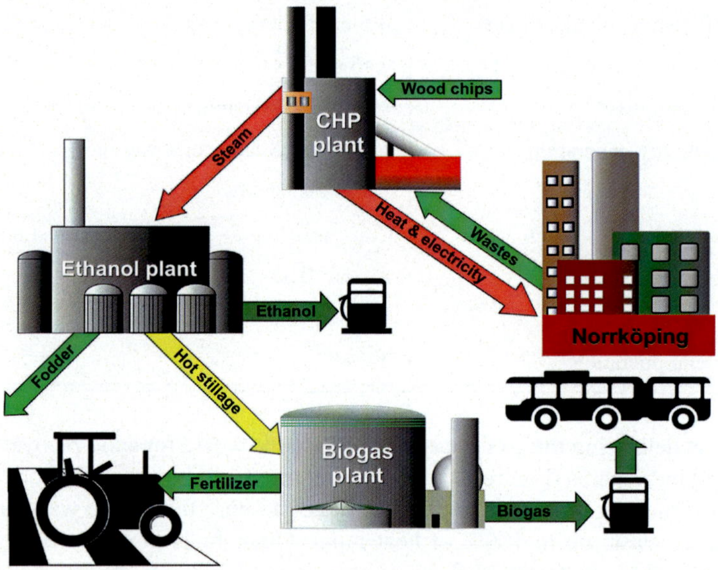

Figure 11.17 Simplified schematic of the energy combine in Norrköping (Martin, 2013).

In addition to ethanol, the plant produces protein for animal feed and renewable carbon dioxide for use in the food industry. The residues that cannot be used for higher value products is fed to a biogas plant together with straw producing biogas from anaerobic digestion. The energy combine is discussed in detail in the thesis of Martin (2013).

11.5 R&D needs and future trends in technological development, markets and applications

It can be concluded from the previous that favorable market conditions supported by sufficiently strong policies have been crucial for the extraordinary development of the

Table 11.8 Some data for the energy combine in Norrköping[a]

Parameter	
Ethanol production	
Raw materials	Wheat, barley, and rye complemented with residues and waste from the food industry and supply chain
Raw material requirement	600,000 tons/year
Ethanol production	225,000 m^3/year
Carbon dioxide	200,000 tons/year[b]
Protein for animal feed	200,000 tons/year
Draff for biogas production	7200 tons/year (dry matter)
Combined heat and power plant	
Fuel power input (five steam boilers)	486 MW
Flue gas condensation	36 MW
Nominal electric capacity	129 MW
Fuels	Biomass and waste (95%)

The numbers regarding ethanol production are given for maximum capacity.
[a]Presently (2015) the older production line is not in operation due to market conditions, which reduces the capacity to 75% of the figures given in the table.
[b]100,000 tons is upgraded by AGA and sold to various applications in the food industry.

biomass-to-energy use in Sweden. As part of the policies, focused efforts to support both fundamental research and applied R&D activities have provided the base for the development of the necessary knowledge, competence, and practical know-how as well as strong networks.

A major future trend is the changeover from an economy based on fossil resources to a bio-based economy—i.e., an economy in which (Formas, 2012):

- Sustainable production of biomass enables increased use in different societal sectors to decrease climate effects and use of fossil resources.
- Increased added value of biomass is achieved using less energy and closing loops by recycling nutrients and use of waste energy with optimized ecosystem services.

This means that as the bioeconomy develops, fossil carbon will be replaced by biomass in a number of applications that will lead to a higher integration of heat generation—for instance, in energy combines and biorefineries (the aforementioned case study from Norrköping is a good example of future solutions).

In addition, the use of waste for heat and power production will play a major role in the future as thermal conversion is sometimes the preferred way to handle waste. Indeed, in a future bioeconomy all combustible waste will be green, and the last use of biomass in a life-cycle perspective will be the production of heat.

The future overall R&D needs defined in Formas's bioeconomy strategy (Formas, 2012) are:

- Replacement of fossil raw materials with bio-based raw materials.
- More intelligent products and use of biomass.
- Changed consumer behavior and attitudes.
- Prioritizing and choice of action.

In the energy sector, the main CO_2-reduction potential in Sweden is within the transportation sector in which the emissions according to a recent investigation (SOU, 2014) can be reduced by 80% by 2030. The investigation is quite comprehensive with detailed analyses and includes a number of proposals for new and revised policies to promote energy efficiency and green fuels.

In the energy system, excluding transport, there is still some potential for replacing fossil fuel with biomass in industrial heating applications. The increased CO_2 tax in industry in 2015 (Figure 11.4) will help to catalyze this development. In district heating and CHP, R&D is, for instance, needed to upgrade and refine fuels and improve fuel flexibility with simultaneous increased technical and environmental performance. There is also a significant potential for increased biopower, which can be an important part of the solution in phasing out nuclear power in Sweden. Furthermore, system perspective and system integration issues will be increasingly important as the biomass sources will be more diverse with increased global trading and as concepts for biorefineries and energy combines develop.

11.6 Ongoing R&D programs

The most relevant research program regarding renewable heating and cooling is the biomass fuel program funded by the Swedish Energy Agency.[5] The quantitative objective with the research is by 2020 (cf. 2009) to achieve a 36–38 TWh increase in sustainable and economically viable production of biomass fuels from the forest (30 TWh) and agriculture (6–8 TWh) sectors. The biomass fuel program is divided into three integrated subprograms: supply, conversion, and sustainability, with the following objectives and contents:

- Supply
  - The major overall objective being to develop biological and technical production processes of biomass fuels to ensure a balance between supply and demand at a reasonable cost fulfilling national and international sustainability criteria. This means, for instance, studies of
    - Effective forest fuel systems—increased supply and development of higher quality fuels at lower cost.
    - Energy crops from agriculture—production methods and technology for the fuel chain including logistics.
    - Forest management for increased fuel production.

[5] A new 4-year program is about to start (second half of 2015).

- Conversion
  - A major objective is to achieve an enhanced competitiveness for the fuel chain from raw material to final use and to increase the raw material base. This includes studies of
    - Refinement of biomass raw material, large-scale pellet production including straw and other residuals from agriculture as well as small-scale refinement adopted to local/regional conditions.
    - CHP—Combined heat and power at small and medium scale (<10 MW).
    - Increased competitiveness of small-scale district heating when replacing oil with biomass.
- Sustainability, including the topics
  - Sustainable biomass fuel production.
  - Effect on climate from production and harvest of biomass fuels.
  - Development of resource and climate effective systems.
  - Effect of policies.

Important applied research programs, with major support from industry, are carried out within Energiforsk (www.energiforsk.se), supporting technical and interdisciplinary research on renewable heat, cooling, and power production from solid, liquid, and gaseous fuels as well as district heating and cooling. Within waste management, Avfall Sverige (www.avfallsverige.se), the Swedish waste management and recycling association, supports applied research.

An important applied research program and network in waste management is Waste Refinery (www.wasterefinery.se). To strengthen the innovation in this area a larger strategic innovation program (ReSource) will be initiated in 2015 (Vinnova, 2015).

Some other academic research centers and programs of relevance include:

- Bioinnovation (www.bioinnovation.se) is a strategic innovation program initiated in 2014. In the program, major stakeholders from industry, academia, and government are cooperating to achieve the vision of a Swedish bio-based economy in 2050.
- Bio4Energy (http://bio4energy.se/), providing a biorefinery research environment. The work is organized into seven platforms: pretreatment and fractionation, thermochemical, feedstock, biochemical, environmental, catalysis, separation, and process integration.
- Chalmers Energy Initiative, in which one focus is on future energy combines (www.chalmers.se).
- Swedish Center for Biomass Gasification (http://www.ltu.se/centres/Svenskt-forgasningscentrum-SFC?l=en) performing research on entrained flow gasification, direct fluidized bed gasification, and dual-bed gasification as well as pretreatment of fuels and gas cleaning.
- Luleå University of Technology (LTU) Green Fuels (http://www.ltu.se/org/tvm/Avdelningar/LTU-Green-Fuels?l=en) in which pilot-scale research and demonstration is carried out for gasification and biofuel production (dimethyl ether and methanol).
- The high-temperature corrosion center at Chalmers (www.htc.chalmers.se/) in which research is carried out to enable enhanced steam data.
- Swedish Knowledge Center for Renewable Transportation Fuels, http://www.f3centre.se/, a national research platform focusing on system-oriented research regarding fossil-free fuels.

Sources of further information and advice

The Swedish development in bioenergy has previously been described (Andersson, 2012). The major public energy research is administrated by the Swedish

Energy Agency, which is also responsible for Swedish energy statistics (www.energimyndigheten.se). Other important government agencies supporting research in the bioenergy field are Formas (www.formas.se), which, for instance, funds research in the field of bioeconomy, and Vinnova (www.vinnova.se), which focuses on the innovation aspects. The Thermal Engineering Research Institute, Värmeforsk (now part of Energiforsk), has for more than 45 years initiated and administrated applied research in the field in close cooperation with industry, and results are provided in a report database (www.varmeforsk.se). The Swedish Bioenergy Association, Svebio, provides a hub for the business stakeholders and policy makers (www.svebio.se).

References

Andersson, K., 2012. Bioenergy-The Swedish Experience. Svebio. Available from: http://www.svebio.se/.

Andersson, K., 2014. Personal communication.

Åmand, L.-E., Lekner, B., Eskilsson, D., Tullin, C., 2006a. Deposits on heat transfer tubes during co-combustion of biofuels and sewage sludge. Fuel 85 (10–11), 1313–1322.

Åmand, L.-E., Lekner, B., Eskilsson, D., Tullin, C., 2006b. Ash deposition on heat transfer tubes during combustion of demolition wood. Energy and Fuels 20 (3), 1001–1007.

Avfall Sverige, 2014. http://www.avfallsverige.se/miljoeteknikexport/presentationer/.

Bioinnovation, 2014. http://www.bioinnovation.se/.

Cronstedt, C.J., 1767. A Publication on Tile Stoves (Beskrifning på ny inrättning af kakelugnar til weds besparning. Jämte bifogade kopparstycken. Utgifwen af præsidenten grefwe C.J. Cronstedt), Libris, 2419358 (in Swedish).

Egnell, G., Börjesson, P., 2012. Theoretical versus Market Available Supply of Biomass for Energy from Long Rotation Forestry and Agriculture – Swedish Experiences. IEA Bioenergy task 43, technical report 2012:02.

Eliasson, D., 2014. Personal communication.

Energymyndigheten, 2011. Fuel Supply Programme 2011–2015. Available from: www.energimyndigheten.se.

Forest Industries, 2013. The Swedish Forest Industries Facts and Figures 2013. Available from: www.forestindustries.se.

Formas, 2012. A Strategy for Research and Innovation for a Biobased Economy. Report R2, ISBN 978-91-540-6067-2.

Jernforsen, 2014. Information available from: www.jernforsen.com.

KSLA, 2013. The Journal of the Royal Swedish Academy of Agriculture and Forestry, Renewable energy – Sweden's unknown green revolution (in Swedish), no. 8, vol. 152.

Lundmark, T., Bergh, J., Hofer, P., Lundström, A., Nordin, A., Bishnu, C.P., Sathre, R., Taverna, R., Werner, F., 2014. Potential roles of Swedish forestry in the context of climate change mitigation. Forests 5 (4), pp. 557–558.

Luomaharju, T., Viljanen, J., 2013. Metso CFB waste-to energy solutions. In: Proceedings Fourteenth International Waste Management and Landfill Symposium. CISA Publisher, Italy.

M. Martin, 2013, Industrial Symbiosis in the Biofuel Industry: Quantification of the Environmental Performance and Identification of Synergies (Ph.D. thesis). Dissertation no. 1507, University of Linköping.

Pelletsförbundet, 2015. Pellet Statistics. Available from: www.pelletsforbundet.se/.
SLU, 2015. Forest statistics from the Swedish National Forest Inventory. Available from: http://www.slu.se/en/webbtjanster-miljoanalys/forest-statistics/area11/.
SOU, 2002. Swedish Governmental Official Report, SOU 2002:100 (in Swedish). Available from: http://www.government.se/sb/d/9682/a/90214.
SOU, 2007. Swedish Governmental Official Report, SOU 2007:36 (in Swedish). Available from: http://www.government.se/sb/d/9682/a/90214.
SOU, 2014. Investigation of a Fossil Free Vehicle Fleet in 2030. Swedish Governmental Official Report, SOU 2014:84 (in Swedish). Available from: http://www.government.se/sb/d/9682/a/90214.
SP, 2015. Information about the P-marking System. Available from: http://www.sp.se/en/index/services/pellets/Sidor/default.aspx.
Svebio, 2015. Available from: www.svebio.se.
Swedish Energy Agency, 2014. Production and Use of Biogas 2013. Report ES 2014 (in Swedish), Available from: http://www.energimyndigheten.se/Global/Statistik/officiell%20statistik/Produktion%20och%20anv%c3%a4ndning%20biogas%202013.pdf.
Swedish Energy Agency, 2010. Proposal for a Transsectorial Biogas Strategy. Report ER, 23, 2010 (in Swedish). Available from: http://www.energimyndigheten.se/Global/Press/Pressmeddelanden/Biogas_Slutrapport_final30aug2.pdf.
Swedish Forest Agency, 2014. Table 7.02 in Statistics and Tables. Available from: www.skogsstyrelsen.se.
The Pellet Handbook, 2010. The Production and Utilisation of Biomass Pellets. Routledge. In: Thek, G., Obernberger, I. (Eds.).
Vinnova, 2015. Strategic Innovation Programme. Re:Source; Available from: http://www.vinnova.se/sv/Var-verksamhet/Gransoverskridande-samverkan/Samverkansprogram/Strategiska-innovationsomraden/Strategiska-innovationsprogram/RESource-/.
Värmeforsk, 2014. Värmeforsk report data base. Available from: http://www.varmeforsk.se/reports.
Waste Refinery, 2013. Perspectives on Future Waste Treatment — Waste Management in Sweden 2013 and Onwards. Waste Refinery Report. http://www.wasterefinery.se/sv/project/projects/perspektivpaframtidaavfallsbehandling/Documents/PFA_eng_130703.pdf.

Index

Note: Page numbers with "f" and "t" denote figures and tables, respectively.

Printed in the United States
By Bookmasters